Philosophy for Heroes
Part III: Act

PHILOSOPHY FOR HEROES

PART III: ACT

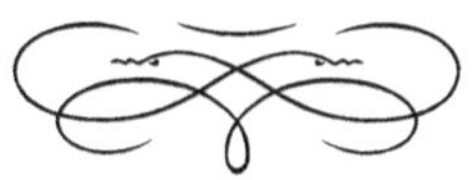

Published by Clemens Lode Verlag e.K., Düsseldorf

NEUROSCIENCE
PHILOSOPHY
POPULAR SCIENCE
PSYCHOLOGY

Dedication

I drew my motivation to write this book from something the neuroscientist Michael Graziano said:

> You go to a conference on consciousness, and in some ways, I feel like I'm at a conference on Tolkien's Middle Earth. And everyone is talking about Middle Earth. Are Orcs really Elves that got modified and this and that and you can talk and talk about it and what it is really magic in Middle Earth and then I'm the iconoclast who is foolish enough to say yeah, that's true, but that's all a description of something and it exists only in simulation. It exists only as information, only as a description. There actually isn't a real Middle Earth. Well, you could say it's a description of a real thing because it's a kind of warped description of Medieval Europe, but it doesn't exist physically as such, and this is very much what I would say about awareness. —*Michael Graziano, Closer to Truth, 2015*

Figure 1: "The Doors of Durin, Lord of Moria. Speak 'friend' and Enter!" —*Lord of the Rings*

Introduction

In this book, I will examine, from the ground up, questions about consciousness. Many steps toward the understanding of the self will tell you nothing about the self—until your right hemisphere connects everything into one idea as you understand the concept. Such an insight is also called an *epiphany*. Using a brain scanner, we can actually observe someone having an epiphany when the brain's right hemisphere suddenly buzzes with activity. While the left hemisphere deals with concrete entities, the right hemisphere helps with looking for alternative meanings. For example, the left hemisphere might identify a "bank" as a financial institution, while the right hemisphere also considers it to be the edge of a river ("river bank").

In the Old Indo-Aryan language Sanskrit, an *epiphany* leading you to the answer about who you are is called "bodhi," which literally means "awakening" or "enlightenment." Similarly, the name "Buddha" means the "Awakened One" or the "Enlightened One." A similar idea can be found in Zen Buddhism as *Satori* (Japanese 悟り) which corresponds to a very *sudden* insight.

This book shows some of the steps leading to Satori, combining the insights of philosophers and scientists into a new idea of what the "self" means. With this knowledge, we can better reflect on our own values and *act* according to reality rather than just blindly following someone else's beliefs. The book series *Philosophy for Heroes* offers the intellectual and moral know-how necessary to be a leader who neither has followers, nor follows other leaders.

Contents

Introduction — vii

Publisher's Note — xi

Preface — xv

5 The Brain — 1
- 5.1 The Evolution of Attention — 4
  - 5.1.1 Nerve Nets in Hydras — 8
  - 5.1.2 Classification of Signals in Arthropods — 9
  - 5.1.3 The Olfactory System in Lancelets — 11
  - 5.1.4 The Optic Tectum in Fish — 15
  - 5.1.5 The Thalamus in Fish — 16
  - 5.1.6 The Basal Ganglia in Fish — 17
  - 5.1.7 The Amygdala in Fish — 18
  - 5.1.8 The Hippocampus in Fish — 20
  - 5.1.9 The Cerebellum in Sharks — 22
  - 5.1.10 The Neocortex in Mammals — 23
- 5.2 The Rise of the Primates — 24
  - 5.2.1 The Primary Motor Cortex — 26
  - 5.2.2 The Rise of the Hominids — 29
  - 5.2.3 The Rise of the Humans — 33
- 5.3 A Glimpse into the Brain — 42
  - 5.3.1 Color Perception — 44
  - 5.3.2 The Brain as a Prediction Machine — 49
  - 5.3.3 The Ventral and Dorsal Streams — 50
  - 5.3.4 The Temporal Lobes — 53
  - 5.3.5 Facial Recognition — 55
  - 5.3.6 Delusions — 58
- 5.4 Creating a Body Schema — 62
  - 5.4.1 The Parietal Lobe — 63
  - 5.4.2 The Premotor Cortex — 67
  - 5.4.3 Alien Hand Syndrome — 70
  - 5.4.4 Phantom Limb Pain — 73

5.5 The Theory of Mind 76
    5.5.1 The Role of the Prefrontal Cortex 78
    5.5.2 Mirror Test 84
    5.5.3 Empathy 88

6 Consciousness **95**
6.1 Theories of Consciousness 101
    6.1.1 History of Theories of Consciousness 102
    6.1.2 The Origin of Consciousness 106
    6.1.3 Monism 107
    6.1.4 Quantum Mind Hypothesis 111
    6.1.5 A Janitor's Dream 115
6.2 The Search for Consciousness 119
    6.2.1 Emergent Property 121
    6.2.2 Blindsight 124
    6.2.3 Split-Brain Syndrome 127
    6.2.4 Hemispatial Neglect 133
6.3 The Loop of Consciousness 138
    6.3.1 The Process of Consciousness 141
    6.3.2 Working Memory 143
    6.3.3 Stream of Consciousness 148
    6.3.4 The Executive Functions 152
    6.3.5 The Inner Voice and Mind's Eye 155
6.4 Model-Building 160
    6.4.1 Supervised Learning 162
    6.4.2 Turing Test 168
    6.4.3 Unsupervised Learning 171
    6.4.4 Summary 176
6.5 The Attention and Awareness Schema Theories . . . 179
    6.5.1 Virtual Prefrontal Cortex 183
    6.5.2 The Awareness Schema Theory 185
    6.5.3 Levels of Consciousness 191
    6.5.4 Awareness Schema and Language 195
    6.5.5 The Subjective Experience 197
    6.5.6 The Limitations that Make Us Human . . . 200

6.6   Building a Conscious Mind 205

      6.6.1   Unconscious Strategies 205

      6.6.2   Conscious Strategies 213

The Book Series *Philosophy for Heroes*   223

Recommended Reading   227

The Author   229

Glossary   231

Bibliography   239

An Important Final Note   257

Publisher's Note

Writing and editing this third book of the series turned out to be a major challenge. Not only did it lead me to change my views on free will and consciousness, but I also discovered it will take more time to put my thoughts into words. That is the reason I split the topics of free will and consciousness into two books. This book will focus on the brain and how it creates consciousness, and the next book will focus on free will and ethics. Consequently, there will be a fifth book focusing on leadership, religion, and heroism.

That being said, the actual process of creating this book has been very rewarding. With the help of the book template (as discussed in *Better Books with LaTeX the Agile Way*), I was able to speed up the writing process and am very pleased about the result. What helped me also was reaching into the community by starting a small local Meetup group and receiving a lot of feedback.

Thank you for keeping up the tradition of reading books. You and your fellow readers have created a market for this book. I hope that I can meet your expectations and I am looking forward to feedback, no matter whether it is positive or negative. To send general feedback, mention the book title in the subject of your message and simply send it to feedback@lode.de. You can also contact us at https://www.lode.de/contact if you are having a problem with any aspect of the book, and we will do our best to address it. Also, we cordially invite you to join our network at https://www.lode.de.

Although I have taken every care to ensure the accuracy of the content, mistakes do happen. If you find an error in this book, I would be grateful if you would report it to me. By doing so, you can help me to improve subsequent versions of this book and maybe save future readers from frustration. If you find any errata, please report them by visiting https://www.lode.de/errata, selecting the book title, and entering the details. Once verified, the errata will be uploaded to our website. You will, of course, be credited if you wish.

Preface

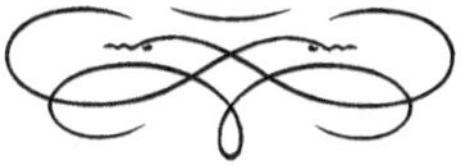

We are not always what we seem, and hardly ever what we dream.

—Peter S. Beagle, *The Last Unicorn*

Imagine one of our ancestors in the distant past, sitting near a lake, lost in thought. She looks into the water and sees her reflection. Then, for the first time in the evolutionary history of humans, the question is asked, "What is this experience I have of myself?" Unbeknownst to her, that question would vex humanity through modern times. Today, there is again an entity looking into a proverbial lake and examining its reflection: while artificial intelligence is still in its infancy, it is on the verge of recognizing itself and asking the same question, "Who am I?"

The closer we come to a machine that seems to be as intelligent as a human being, the more we start to worry about our own subjective experience. If a computer eventually becomes indistinguishable from us, what makes humans special? What is our role in the universe if we are so similar to a computer program? *Does your brain need a self at all?*

My goal with the *Philosophy for Heroes* series is to give you a comprehensive overview from the very fundamental ideas of ontology ("What is?") and epistemology ("How do we know?") through the fields of psychology, ethics, and leadership. By being able to reflect on who we are, we can become better leaders.

Sitting on my balcony and watching the stars in the clear sky, I am grateful for the past 10 years I spent in research and emotional development. I see how people are struggling to cope with the current crisis, be it with anxiety, hate, or outright denial. Reality is tough as it can be unforgiving, without a political agenda, and final. In a way, studying science is a coping mechanism, too. If I know how scary things work, they lose part of their power. If I understand how A will lead to B, I do not have to hide under the bed because B might jump through the door at any moment. I can look for evidence for A or try to evade or prepare for it.

What I wrote five years ago in *Philosophy for Heroes: Knowledge* remains true:

> I hope that you will take with you from this book series one or two exciting ideas, and develop them further or let them inspire you. Personally, I would like to place this book in the hands of a younger version of myself, someone who finds him- or herself at the beginning of his or her journey of scientific discovery, having to first wade through tons of misinformation before getting to the "good stuff." Even if I reach only a small handful of people who will take to heart a few of these core ideas and set out to achieve something great in life themselves—whatever that might be—I will be able to enjoy the rewards of this book. I made the book available to the public just as I would plant a handful of seeds in the earth, hoping that they grow and bloom.

In *Philosophy for Heroes: Knowledge*, we started our journey into philosophy with the intuitive understanding that the world consists of entities which we can objectively perceive. For this, we need to keep going back and forth between information we gather about the world (*ontology*) and about how we perceive the world (*epistemology*). Whenever we discover something new, we should re-examine how this affects our knowledge and perception of the world. For example, if we find out that someone is a liar, we should re-examine what we learned from that person.

In *Philosophy for Heroes: Continuum*, we formalized this approach with the scientific method. Using this tool, we explored ontological questions in physics and biology to answer questions about the origin of the universe and life.

In *Philosophy for Heroes: Act*, we will continue with the scientific approach to explore the evolutionary history and architecture of the brain. Using these new insights, we will draw conclusions on how we experience the world and create a sense of self-consciousness.

You, the reader, are holding the open book in your hands and now have an overview of what to expect.

Let us continue our journey to become a shining example for others, thus getting closer to our ideal world. The path to our goal is the *Philosophy for Heroes.*

Clemens Lode
Düsseldorf, Germany, November 1st, 2020

You can find chapters 1 and 2 in

PHILOSOPHY FOR HEROES

PART I: KNOWLEDGE

Published by Clemens Lode Verlag e.K., Düsseldorf

You can get a free copy at https://www.philosophy-for-heroes.com

You can find chapters 3 and 4 in

PHILOSOPHY FOR HEROES

PART II: CONTINUUM

Published by Clemens Lode Verlag e.K., Düsseldorf

Chapter 5

The Brain

> Our imagination is stretched to the utmost, not, as
> in fiction, to imagine things which are not really
> there, but just to comprehend those things which
> are there.

—Richard P. Feynman, *Character of Physical Law*

When we look at a picture or a scene, we take for granted that what we are seeing is the entire picture, that we have an objective view, and that what we see is real. However, we are not cameras. We *think* we have looked at the whole picture or scene, but, in reality, we likely have perceived only a filtered version of it, and all we might remember are several details that caught our attention. To explore this curious difference between our experience and reality, the first step is to understand the brain's architecture. Given that consciousness must sit somewhere between perception and action, we will examine the entire stream of data, from sense data to action.

The Evolution of Attention. Before we can discuss the brain's higher functions, we first have to understand all its parts. The best approach is to look at its evolution from the first multi-cellular life-forms and work our way up to the brain of modern mammals.

The Rise of the Primates. To comprehend the most modern part of our brain, the neocortex, we have to go into depth and compare our brain to that of our closest cousins, the apes. Then, to gain insight into how our brain evolved, we also need to take a look at the environments in which our human ancestors spent most of their evolutionary history.

A Glimpse into the Brain. To learn about our subjective experience of the world, we need to look at how sense data arrive in and are processed by our brain. For this, we examine the visual system from the eyes to the visual cortex.

Creating a Body Schema. To understand how our brain initiates actions, we need to know how the brain gains a sense of our body. We build a body schema to be able to differentiate between ourselves and our environment.

Theory of Mind. Finally, we need to explain how the brain makes sense of itself. Consciousness goes beyond seeing oneself in the mirror; it also involves knowing what we (and other people) are thinking about.

5.1 The Evolution of Attention

What evolutionary steps contributed to the development of the conscious experience and process of decision-making humans possess today?

Figure 5.1: What we most closely connect to someone's attention are his eyes. Not only can the eyes tell us what someone is looking at, but they can also give us hints about what someone might be thinking (image source: Shutterstock).

ATTENTION · *Attention* is the brain's process of limiting alternative thought patterns, then increasing the most dominant thought pattern's strength. It is like a simple majority rule: the most successful thought pattern gets all the resources while other thought patterns are suppressed. While we can jump back and forth between different thoughts, we cannot have two dominant thought patterns at the same time.

Attention refers to the ability to select between competing or even contradicting sense data. For example, you hear something on your left, see something moving on your right, you are hungry, and tired; which sense data gets your attention first? The brain parts involved that help us make such a decision underwent half a billion years of evolution and can be traced back to simple multi-cellular organisms.[1] Figure 5.2 shows the evolutionary timeline of primates with different species branching off. As we are looking only at the evolution of attention, our focus will be a small selection of species rather than a comprehensive discussion. Figure 5.3 shows the same tree of dependencies in a graphical form, with branches representing the creation of new major species.

Time	Species	Brain part
600 mya	sponges	calcium signalling
580 mya	*Hydras*	basic nerve net
550 mya	arthropods	information classification
535 mya	lancelets	olfactory system
520 mya	fish	optic tectum (information tracking)
520 mya	fish	thalamus (information integration)
520 mya	fish	basal ganglia (resolving conflicts)
520 mya	fish	amygdala (information evaluation)
520 mya	fish	hippocampus (spatiotemporal memory)
450 mya	sharks	cerebellum (movement programs)
300 mya	reptilians	wulst (high-level processing)
225 mya	mammals	neocortex (similar to wulst)
55 mya	primates	prefrontal cortex (planning)

Figure 5.2: Timeline (in million years ago) of the evolution of attention in animals with a list of species that branched off the evolutionary tree and the brain part that first appeared at that time.

---

[1] Kaas, 2017, p. 547–554.

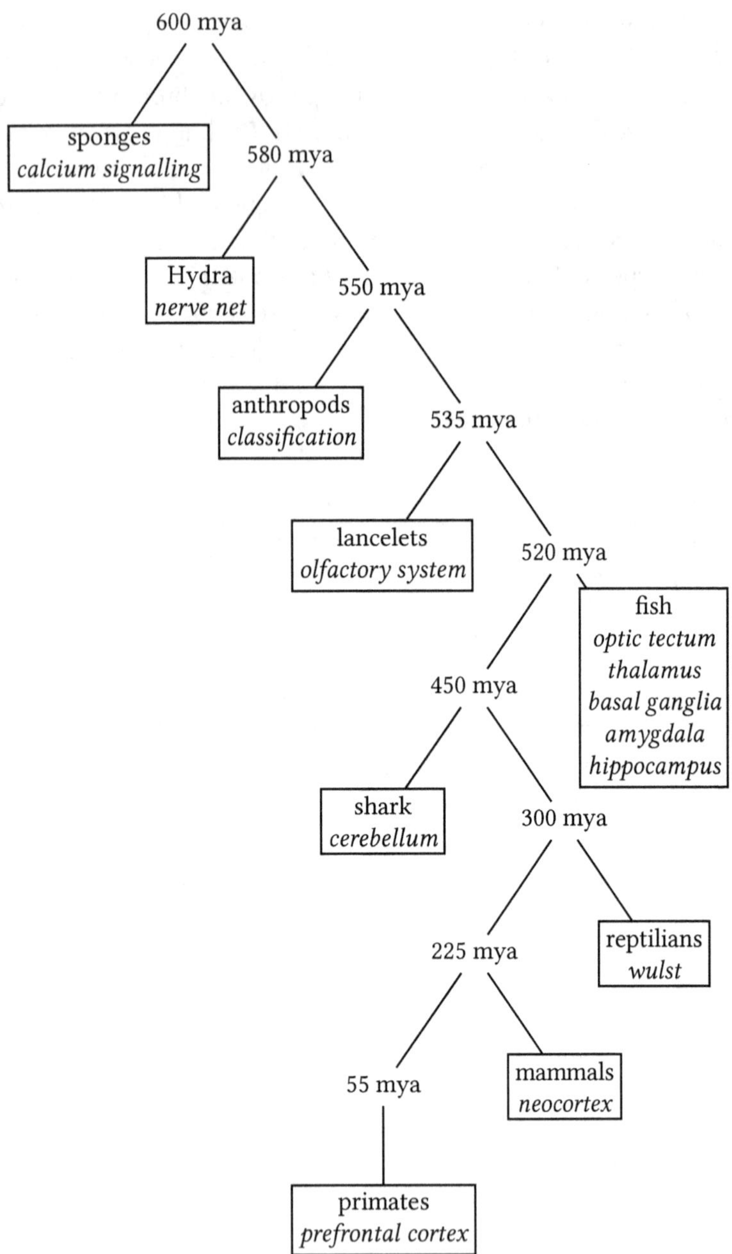

Figure 5.3: Simplified evolution of attention in animals.

Over this long history, the brain became a collection of different functions, layered on top of each other, with many systems having overlapping responsibilities. If a neural pathway helped one of your ancestors to avoid danger, that pathway survived, even if that meant that the architecture became somewhat chaotic ("complex").

Nature does not care about an easy-to-understand architecture; it cares only about what works and what does not.

Sometimes, the organization of a certain function into clearly distinct brain parts had an evolutionary advantage (for example, separating the *neocortex* from the rest of the brain). In other cases, the most efficient layout was having one function directly beside the other (for example, the different brain regions within the necortex). Yet, for our understanding of the brain, it is sufficient to look at it as a system of separate parts interacting with each other. We just have to keep in mind that the functions of brain parts usually blend into those of neighboring brain parts, and that there are many more interactions and connections between brain parts than listed here.

Major structural components and properties of the brain include:

ALLOCORTEX · The *allocortex* is part of the cerebral cortex (the *neocortex* is the other part) and consists of the olfactory system and the hippocampus.

NEOCORTEX · The *neocortex* is the newest part of the mammalian brain and consists of the *cerebral hemispheres*. Its main tasks are focus, language, long-term planning, and modelling of the world. It can generate strategies that involve detours if goal-directed behavior is not successful (for example, going around a fence instead of trying to get through it).

CEREBRUM · The *cerebrum* includes the neocortex (the cerebral hemispheres), and the allocortex (the hippocampus, the basal ganglia, and the olfactory bulb).

| **CEREBRAL CORTEX** · The *cerebral cortex* is the outer layer of the cerebrum. It contains most of the neurons of the brain.

| **GYRUS** · A *gyrus* is a fold or ridge in the cerebral cortex.

| **SULCUS** · A *sulcus* is a groove in the cerebral cortex.

5.1.1 Nerve Nets in Hydras

Looking at our tree of animal ancestors (see Figure 5.3) in regard to brain development, sponges were the first to settle into an evolutionary niche (more than 600 million years ago). They are sea animals that are mostly immobile and simply filter oxygen and nutrients from the ocean water. Although they have a primitive way of pushing water, when it is toxic or otherwise polluted, out of their bodies, they lack any form of nervous system as we understand it. Their cells communicate directly with each other using calcium signalling. Each cell contains a concentration of calcium that can be released if it receives calcium from neighboring cells. This way, it creates a calcium wave propagating throughout the organism. You can imagine it like having many square containers grouped together and filled to the brink with water. When you take one container and pour it into its neighboring containers, all the containers will overflow.

The first animals with some semblance of a brain were *Hydras*. They branched off our evolutionary tree more than 580 million years ago. They are small (around 10 millimeters in length) animals that usually attach themselves to the surface of an object in their environment and can slowly move over it or detach themselves and float in the water. They have a basic nervous system that allows them to use their tentacles to attack prey. If another animal (mostly tiny planktonic crustaceans like *Daphnia* or *Cyclops* up to five millimeters in length) touches a tentacle, the nerve cells activate the tentacle to take that animal into the *Hydra*'s mouth. There is no central nervous system that organizes this activation. Instead, nerve cells are spread throughout the body of the *Hydra* in a nerve net. This enables the *Hydra* to respond to its environment without being able to detect where this original stimulus came from. Any signal leads to the same reaction—for example, all muscles contract at the same time.

If the human body had a nerve net instead of a nervous system and brain, we would not be able to figure out where we were touched, only that we felt *something* and as a result had to come up with a general response to this touch. Such a general response is comparable to our hormonal system: for example, the adrenaline released in a situation of danger does not cause specific actions but prepares the whole body for a possible injury or energy exertion. Another example would be the regulation of body temperature which, again, is a general response to certain conditions instead of a specific movement.

5.1.2 Classification of Signals in Arthropods

A basic form of attention appeared at the time the arthropods (insects, spiders, crabs, etc.) split off the evolutionary tree around 550 million years ago. With this new form of attention, instead of treating all sensory input as equal, the information is pre-processed and can thus be amplified and classified. Imagine noticing something suddenly moving in the grass—it immediately draws your attention. Once you see it emerging from the grass, you classify it as a particular concept, for example, a snake.

At its core, classification is about *filtering* information we do not need. No longer would every signal cause a reaction. Instead, the organism was able to focus on specific signals and react to those.

Dog at 1pm	Dog at 2pm	Analysis
Basket	Basket	Dog has not moved
Rug	Rug	Dog has not moved
Basket	Rug	Dog has moved
Rug	Basket	Dog has moved

Figure 5.4: Example for an application of the XOR filter. If the dog is at different places at 1pm and 2pm, we can conclude that the dog has moved.

Most multi-layered nervous systems (including our own) support this kind of filtering. By comparing several images on your retina for changes, your visual system can make out which moving part belongs to which previously seen part. For example, if your visual system identifies a dog and then the same dog in subsequent images, you perceive any changes in those images as movements of the dog. If you closed your eyes every second, you would perceive the dog "jumping" from place to place. You would have to use your short-term memory as a workaround and remember where the dog was earlier to decide whether or not he had moved. Figure 5.4 shows an example that represents the function *XOR* ("eXclusive OR") which filters out similarities and returns differences. When passing two images through such an XOR filter, it would highlight changes between both images and thus detect movement.

Visual pre-processing is done partly by the retina of our eyes, detecting edges and changes, and compressing the data-stream toward the rest of the visual system. To understand what is happening, imagine looking at a picture of a palm tree in front of a white background. The brain perceives the detailed raw image, then the visual system extracts the edges of objects to identify them (see Figure 5.5). This way, the brain can determine that there is the shape of a palm tree. This is the opposite of what happens when *drawing* a picture: we *start out* with the palm tree in mind, then draw the edges and contours and then finally fill them in with details.

Figure 5.5: Edge detection applied to the image of a palm tree (image source: Shutterstock).

Beyond being just a one-way street of information (classifying the image data to abstract information), classification systems can also help you to direct attention. When we see things that are new or unusual, our brain allocates resources to finding out what they are. This could play out by turning our head, refocusing our eyes, looking at things from a different perspective, going closer, or asking others about the new or unusual things.

5.1.3 The Olfactory System in Lancelets

The olfactory system was probably one of the earliest sense organs that evolved in animals, as detecting molecules is closely connected to a lifeform's search for nutrients. While not directly related to us (they diverged from our ancestors around 535 million years ago), we share some of our olfactory-related genes with lancelets (see Figure 5.6). They can be seen as predecessors of fish with similar organs

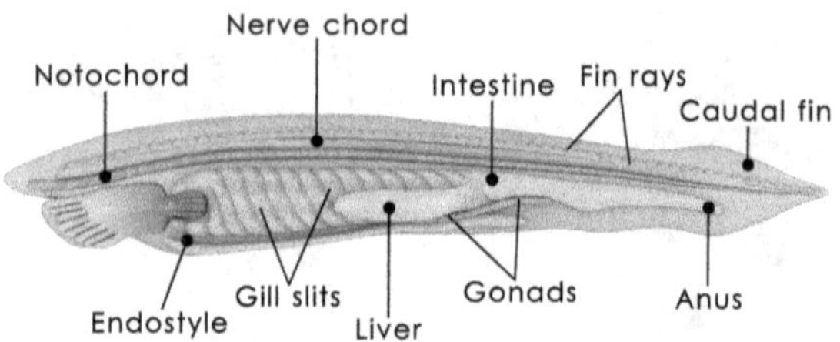

Figure 5.6: Lancelets can be seen as predecessors of fish with similar but more primitive organs (image source: Shutterstock).

but in more primitive form. For example, their gill-slits are used for feeding but not for respiration. Likewise, their circulatory system transports nutrients but not oxygen. While they have no centralized olfactory system (we humans do), their olfactory receptors are studded along their flanks to detect possible sources of nutrition in their aquatic environment.

Molecules connect to the olfactory system over the peripheral olfactory system. In aquatic animals, this happens directly via contact with the water. In land animals with lungs, this happens by having the airborne molecules dissolve into mucus on top of the olfactory receptor cells. If the molecule binds to the receptor cell, a nerve signal is created and transmitted to the brain. A peculiarity of our sense of smell is that it is the only sense that can bypass the thalamus (see Chapter 5.1.5) and send signals directly to the neocortex.

While our sense of smell might seem to play little to no role in our modern hectic life, it actually has a significant impact. In combination with our sweat, our sense of smell can communicate emotions. Usually, we think of emotions being contagious by way of our sense of sight or sense of hearing—we tend to laugh when we see or hear someone else laugh. But studies have shown that emotions are also contagious via our sense of smell. This works even when the smell is separate from the person (for example, on his clothes). So, even without words or gestures, people can communicate their distress to others nearby.[2] The evolutionary advantage of this mechanism makes sense, especially when it comes to fear. Putting yourself into a heightened state of alertness when detecting fear in other people can increase your chances of survival.

The olfactory system also supports mate selection by detecting pheromones which contain the MHC complex. In *Philosophy for Heroes: Continuum,* we discussed the relevancy of the MHC complex in the immune system's ability to differentiate self from other. In mate selection, a similar system is used to find a partner that is genetically not too similar but also not too different. The evolutionary advantage is to have a compatible partner with increased resistance to infectious diseases by providing a variability in the MHC complex.[3] At the same time, it reduces the chance for children to inherit genetic diseases. Similar to the immune system, the olfactory system probably becomes accustomed to the MHC complex of relatives in early childhood. If this contact does not happen, there is no biochemical obstacle to falling in love with close relatives.[4]

In terms of brain architecture, information travels not only from the olfactory system to the brain, but also in the opposite direction. If a particular faint smell wins the neural competition, resources are allocated to enhance our olfactory system's sensitivity. This focus can improve the olfactory system's efficiency by providing context information. In fact, the information from the millions of odor detectors in the olfactory system never even arrives at our neocortex. Instead, it is condensed into only 25 cells which are primed by the neocortex. If there is a strong smell, the sensitivity of the cells is reduced; if we want to pick up a faint smell, we can increase the sensitivity.

---

[2] Mujica-Parodi et al., 2009.

[3] Ejsmond, Radwan, and Wilson, 2014.

[4] Potts and Wakeland, 1993.

By combining the gustatory system (the basic tastes sensed by the tongue like salty, sour, bitter, umami, sweet, kokumi, calcium, and so on) with the smells detected by our nose, we can enhance our overall experience of food. Children learn to like or dislike certain types of food when observing what is safe for other people to eat.[5] While individual exceptions exist, if humans were genetically disposed to favor a particular food (like Koala bears prefer eucalyptus tree leaves) to the exclusion of other foods, our ancestors would have had a hard time spreading all over the globe.

If the olfactory system classifies something as inedible, it might initiate the gag reflex to protect the body from poisons. If we actually get food poisoning or an infection, the body reacts by increasing acetate levels in the blood. In the brain, this improves the ability to create memories. The evolutionary advantage of this pathway could be to better remember the situation that led to the food poisoning or infection and thus prevent it in the future.

All these properties are reflected in the architecture of the olfactory system. There are the following connections (Figure 5.7):

- Trigeminal nerve, vagus nerve (gagging reflex, face muscles, expression of disgust);
- Hippocampus (spatial memory);
- Amgydala, hypothalamus (emotional reaction, hormones, pheromone processing);
- Neocortex (processing of smells);
- Hippothalamus (pheromones, hormones);
- Olfactory bulb (sensory cells); and
- Nose (air flow).

---

[5] Elsaesser and Paysan, 2007.

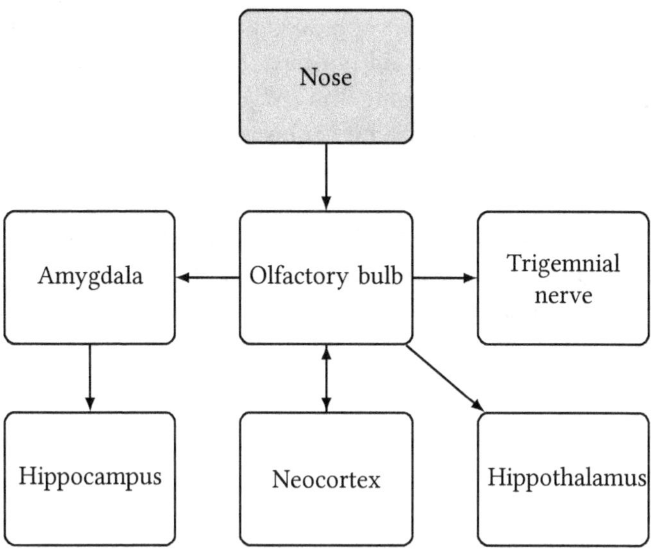

Figure 5.7: The architecture of the olfactory system.

5.1.4 The Optic Tectum in Fish

Controlling eye movements made it necessary for fish (they split from the evolutionary tree around 520 million years ago) to develop central processing, namely the *optic tectum*. In mammals, this organ is called the *superior colliculus* and most of the processing has moved to the visual cortex. It helps fish (and us) to track moving objects and is responsible for blinking as well as pupillary and head-turning reflexes. Relying on auditory information, the superior colliculus is also responsible for reflexively turning one's eyes and head toward a sound source.

In the brain, the superior colliculus sits right behind the optic chiasm where the nerves from the left eye and right eye cross. If something or someone outside of your eyes' focus moves, this part of the brain is responsible for bringing it to your attention. You might then turn your head and re-focus your eyes to get a better picture of the possible threat. Imagine you did not have this reflex to see anything moving in your environment. The risk of injury (say, from an oncoming tiger) would be much higher because of your longer reaction time.

SUPERIOR COLLICULUS · The *superior colliculus* or *optic tectum* (in non-mammals) helps the eyes to track objects, and controls blinking, pupillary, and head-turning reflexes.

5.1.5 The Thalamus in Fish

On the evolutionary timeline, the *thalamus* also first appears in fish. It combines and pre-processes different sources of sensory information into a coherent whole before relaying it to other parts of the brain:

- It combines the information from the left eye and right eye to build a three-dimensional representation of the environment.

- In humans, it translates the signals from the red, green, and blue cone cells in the retinas into colors. While further processing takes place in the neocortex, the first part responsible for this color encoding is the *lateral geniculate nucleus* (or LGN).

- Like the superior colliculus, the LGN also receives auditory information. The LGN changes the auditory information so that you perceive the sound as coming from a visual source. For example, when watching television, the LGN "moves" the perceived location of the source of sound from the speakers to the screen.[6]

---

[6]McAlonan, Cavanaugh, and Wurtz, 2006.

> **THALAMUS** · The *thalamus* integrates different sensory information and relays the information to other brain parts. For example, it combines sense data from the retinas' cones into colors, or calculates three-dimensional information from the two-dimensional images from both eyes.

> **LATERAL GENICULATE NUCLEUS** · The *lateral geniculate nucleus* (LGN) is part of the thalamus and relays information from the retinas (via the optic chiasm) to the visual cortex. It pre-processes some of the information, for example, combining red, green, and blue photoreceptor cells into colors.

5.1.6 The Basal Ganglia in Fish

For the evolutionary competition of neurons (that we have discussed in *Philosophy for Heroes: Continuum*) to work, research points to an involvement of the basal ganglia.[7] The *basal ganglia* are thought to originate from the need to arbitrate between different courses of motor neuron activation. They identify the "best" among several possibly contradicting courses of action. Rules determine what "best" means in a particular context. For example, you can have the two competing thoughts (e.g., wanting to go left and to go right), but you cannot physically walk in two directions at once.

> **BASAL GANGLIA** · The *basal ganglia* are a part of the brain that, like a referee, arbitrate decisions by the neural committees. Also, like an orchestra conductor, they coordinate the sequence of entire motor programs. In both cases, they do not make decisions but merely provide rules and structure.

Beyond selecting individual motor actions, the basal ganglia seem to be involved in cognitive thought patterns. While contradicting thoughts can exist, different thought patterns cannot recruit the

---

[7]Redgrave, Prescott, and Gurney, 1999.

same cognitive resources at the same time. For example, imagining a unicorn leads to thought patterns recruiting parts of your visual cortex. Adding more elements to the scene requires more and more resources until you can no longer focus on all elements at the same time. The basal ganglia are also involved in coordinating entire motor programs. Imagine an orchestra without a conductor: sure, the musicians could play their respective parts, but at different speeds and starting at different points in time. The basal ganglia act like a conductor of an orchestra, synchronizing the different motor programs, activating them in the right sequence and with the right timing.

5.1.7 The Amygdala in Fish

Yet another brain part, the *amygdala*, first appeared in fish. Managing our attention with the basal ganglia is one thing, how we *prioritize* the signal is another. While we can make decisions based on the strength of the signal—turning our head to the loudest noise seems to be a good strategy—we also need to put the signal into context. For example, instead of always running away from a tiger, we might consider whether or not to take the risk and first pick some berries and only then run away—especially if we are very hungry. This demonstrates how the amygdala uses information from a number of sources to prioritize different courses of action.

> **AMYGDALA** · The *amygdala* is the brain's value and emotion center. It helps with evaluating thought patterns of the basal ganglia depending on the context instead of the mere strength of the signal. It also connects the brain with the hippothalamus, providing a bridge to the hormonal system.

With its connection to the hormonal system, the amygdala also initiates fight, flight, freeze, or fawn responses when in distress:

- Attack the predator ("fight" response);
- Run away from the predator ("flight" response);
- Remain still ("freeze" response); or
- Display submissive behavior ("fawn" response, by humans and other social animals).

How the fight and flight responses can help in a threatening situation is self-explanatory. The "freeze" response can trick predators because many predators' instincts depend on motion. If their prey is not moving, the predator's hunting instinct is not activated and they will look elsewhere for food. For example, cats take great interest in a moving toy while they might ignore something that remains still. Similarly, the "fawn" response works if the attacker is from the same species and also a social animal. Showing submissive behavior communicates to the other party that you are not a threat, preventing possible injury for both parties.

Beyond helping with the immediate response (e.g., releasing adrenaline), it can also serve as an early basic form of memory. For the response to be effective, the hormonal changes caused by, for example, the flight response need to remain long after a predator has vanished from an animal's view. It will cause it either to head home to a safe place or to be on alert when it returns to this location.

The information the amygdala is using is limited to immediately available sensory data. It activates emotional reactions based on mapping the input from the thalamus to emotional behavior. For example, the sight of fresh berries might evoke a positive emotional response while the sound of a rival might evoke a fight, flight, or fawn response. This response takes priority over any rational evaluation of the situation because it is quicker and possibly stronger than signals coming from the neocortex. The amygdala associates sense data coming through the thalamus with positive or negative events and, ultimately, emotions, for example:[8] Tiger $\Rightarrow$ fear; apple $\Rightarrow$ appetite; and sun $\Rightarrow$ happiness. What makes this mechanism so powerful is that it requires very little processing power while it can cover a wide range of sensations. The major limitation of behavior based on the amygdala is the limited range of reactions and the reliance on immediately available sense data. The amygdala cannot take into account abstract thinking or planning, or complex relationships between objects, animals, or people.

5.1.8 The Hippocampus in Fish

A predecessor of our *hippocampus* also developed around the time of the first fishes. The actual hippocampus is unique to mammals but there are theories that similar structures evolved from a common ancestor of reptiles and mammals around 520 million years ago. Its main task is to create a mental map of an animal's environment to allow the animal to remember where possible food and water sources are located. It also helps with navigation, remembering paths the animal has taken, relating spatially to other animals or objects,[9] and recognizing places for orientation. A good example for the use of the hippocampus is squirrels burying nuts as food stashes for the winter. Our current understanding is that this map is not a literal map but instead consists of points of orientation. While many people can construct a mental image of a map, we tend to orient ourselves by seeing something we know and then putting our goal in relation to the landmark. For example, when describing to another person the path to a location, we might say "Walk down the street until you get to the large tower, then turn right."

---

[8]Tye et al., 2008; Rogan, Stäubli, and LeDoux, 1997; McKernan and Shinnick-Gallagher, 1997.

[9]Danjo, Toyoizumi, and Fujisawa, 2018.

> **HIPPOCAMPUS** · The brain's *hippocampus* provides us with a mental map for navigation. It also builds temporal relationships between places, allowing us to determine, for example, which areas in our environment we have already foraged and in which areas the plants have regrown. The *hippocampus* and the *olfactory system* (sense of smell) make up the *allocortex*.

While earlier animals could drift and react to sensory inputs (evading predators and approaching food), once the food was out of sight, it was also out of mind. The hippocampus allowed animals to find more food by avoiding areas that they had already foraged and exploring areas they had not foraged. This requires mapping the environment based on odors (the sense of smell has direct connections to the hippocampus) and sights, as well as prioritizing those according to the time they should be visited.[10] This led to the evolution of the hippocampus to handle tasks in serial order with the right timing and in the right context. This stems from its ability to associate two memories with each other, which helps to find a path from one place to another.[11]

Did you know?

The hippocampus' function becomes most visible during dreams, when experiences retained during the day are played back for long-term memory backup in the neocortex. While we cannot ask animals whether or not they dream, some animals show rapid eye movements (REM) in their sleep, pointing to an activation of their hippocampus.

⟶ Read more in *Philosophy for Heroes: Persona*

---

[10]Murray, Wise, and Graham, 2018.
[11]Samsonovich and Ascoli, 2005.

5.1.9 The Cerebellum in Sharks

More than 450 million years ago, sharks with a *cerebellum* emerged. This organ coordinates complex, time-critical behavior which includes movements, speech, and balance. When hunting, the shark might have had to outmaneuver its prey and then bite at the right moment. Similarly, its prey had to come up with movement strategies to navigate through the water to evade predators, locking predator and prey into an evolutionary race. Mammals face similar challenges of coordination when trying to jump from tree to tree, evade attacks by predators, or catch prey. Given that both the cerebellum and the basal ganglia are involved in coordinating motor programs, it is no surprise that they also form an integrated network to exchange information.[12]

> **CEREBELLUM** · The *cerebellum* is the brain part that helps with coordination of complex behavior. It provides a set of motor programs the brain can choose from repeatedly for similar actions (even in time-critical situations). With the help of the cerebellum we can, for example, walk or bicycle without having to consciously think of each movement.

In more general terms, the cerebellum is responsible for replaying movements. This becomes apparent when examining how the cerebellum learns new movements. Think about how each leg moves, as you did when you first learned to walk or to ride a bicycle: the programs to coordinate all your motor neurons were not yet transferred to your cerebellum ("learned"). Thus, they were not yet optimized and consequently, they were very slow. Until that optimization happened, walking or riding a bike had not become "second nature." You had to take "baby steps" and focus on one step or motion at a time before making the next one.

---

[12]Bostan and Strick, 2018.

In the (very rare) case of people born without a cerebellum, we see late walking development, reduced gait speed, unsteady gait, and a reduced ability to stand in darkness or when their eyes are closed.[13] This is explained by the fact that the cerebellum is connected to the inner ear, providing our sense of balance. In addition, people born without a cerebellum have late speech development, slurred and slowed speech, and a reduced control of pitch and loudness. This points to an additional role of the cerebellum in language fluency.[14]

5.1.10 The Neocortex in Mammals

The neocortex, which is layered over the *collicular control* of attention (above the previously mentioned superior colliculus), developed more than 300 million years ago. Following the Permian-Triassic mass extinction event 252 million years ago (extinguishing 70% of land biodiversity), both the dinosaurs and mammals emerged. Given that dinosaurs still dominated the planet, mammals had to move into a niche and become nocturnal animals. According to the *nocturnal bottleneck hypothesis*, traits of growing fur, managing body temperature, and well-developed senses of smell, hearing, and touch helped our ancestors to stay active at night while evading predators during the day.

> **NOCTURNAL BOTTLENECK HYPOTHESIS** · The *nocturnal bottleneck hypothesis* posits that many mammalian traits were adaptions to moving into a niche to become nocturnal animals and evade the dominant dinosaurs.

Functions of the neocortex include sensory perception, conceptualization, directed motor commands, spatial reasoning, communication, and long-term planning. The visual field is analyzed to create a mental representation of objects and their location. Instead of the

---

[13]Yu et al., 2014.
[14]Carta et al., 2019.

raw sense data, this mental representation of the world is then used by the rest of the brain in, for example, coordination with motor control, language (reading this book), or face recognition. Similarly, other senses (hearing, touch, smell, etc.) are processed, and their signals are categorized and prioritized.

While we are now aware of the principal brain functions we share with other mammals, the question remains what makes us humans unique. In Chapter 5.2, we will look at how our human ancestors adapted to the Savannah and how that set them apart from their closest primate cousins, the chimpanzees, that stayed behind in the forest.

5.2 The Rise of the Primates

To understand human nature, we can compare ourselves with our closest primate cousins, the chimpanzees. How do we differ, what makes us special, and what has taken us on two different paths?

Some 66 million years ago, a 10km- to 15km-wide asteroid hit what we today know as the Gulf of Mexico and in its aftermath killed all animals over 55 pounds body weight (the *Cretaceous-Paleogene extinction event*). This opened an opportunity for the smaller mammals to slowly return from their nocturnal life. Prior to this fateful event, mammals had adapted by hiding during the day. At night, they relied on their sense of smell to evade predators and to locate prey. This led them to evolve abilities that turned out to be useful when they returned to the daylight. One of those abilities was suppressing immediate urges. With the help of the *prefrontal cortex*, the mammal was able to save valuable time and energy. It could ignore scents of predators that were long gone, and pursue prey that was still nearby.

> **PREFRONTAL CORTEX** · The *prefrontal cortex* is part of the frontal lobe and can be understood as running a simulation of the world. It monitors social relationships, keeps track of objects when they are no longer visible (object permanence), and helps with the pursuit of long-term goals. It has only indirect connections to brain parts dealing with actions or sense perception.

When mammals later adapted to the daylight, the same principle was applied to visual data. For example, just seeing a lion does not mean that the lion is dangerous; it might be sleeping. To make that judgement call, the separate evaluation and signal by the prefrontal cortex was needed as a counterbalance, thinking long-term, balancing risk and reward.

The prefrontal cortex is only indirectly connected to any sensory organs or muscles, hence it is believed to process thoughts that do not need any external input to be activated and do not necessarily result in a concrete action to be executed. Put differently, the prefrontal cortex reads and controls the rest of the brain instead of directly accessing sense organs or muscles.

To understand the prefrontal cortex' role, we need to remember that many parts of the brain are in competition with each other. In that regard, the prefrontal cortex can be understood as a counterbalance to the rest of the brain. Applications include long-term goals, social rules, hidden threats, imagination, or memories. For example, a strong visual input like a tiger standing in front of us would need an equally strong competing signal to prevent a simple fight, flight, fawn, or freeze reaction.

5.2.1 The Primary Motor Cortex

The first true primate ancestor was a small nocturnal monkey-like, forest-dwelling animal (*Plesiadapis*) that lived in trees and ate fruit (around 55 million years ago). This primate-like mammal needed strength and balance to jump from tree to tree, and a brain that could calculate movement in three-dimensional space, as opposed to just navigating on the ground. Also, its eyesight had to improve to adapt to the light of day and to recognize ripe fruit. A testament to this evolutionary history is that we now lack the ability to produce our own vitamin C and need to rely on a steady diet of fruit and vegetables.

About 30 million years ago, the evolutionary branch split into Old World monkeys and New World monkeys. The most widely held theory about this split is that the ancestors of the New World monkeys used a temporary land bridge or series of islands between Afro-Eurasia and South America. The main difference between the two types of monkeys is that New World monkeys kept their tail. With their *primary motor cortex* (see Chapter 5.4.2) being focused on their tail muscles, they use it like a fifth limb. In contrast, the primary motor cortex of apes (Old World monkeys) is specialized for hand use which helps with foraging for fruit. To observe this in humans, just stand in a modern supermarket and watch people carefully checking out the ripeness of avocados.

> **PRIMARY MOTOR CORTEX** · The *primary motor cortex* is part of the frontal lobe and is directly adjacent to the primary somatosensory cortex of the parietal cortex. This way, it can directly process data from our sense of touch to better control movements. The *primary motor cortex* connects to the *brainstem* and *spinal cord* (via the *upper motor neurons*) which in turn connect to the muscles (via the lower motor neurons).

The feature that makes the primary motor cortex in primates special compared to that of other mammals is that it can bypass the spine's interneurons (the spine's relay station) and send signals directly to the motoneurons of the spine.[15] With a significantly higher number of neurons controlling the movement, this allows fine-grained control of the actual signals sent to the muscles. This part of the brain also grew overproportionally during the evolution of primates.[16]

Why exactly do more neurons lead to better control?

Our primary motor cortex contains about 5 billion neurons controlling some 600 muscles. Those 600 muscles consist of more than 50 billion muscle cells. But muscles can only contract, so, theoretically, to contract all 600 muscles individually, no more than 600 motor neurons would be needed. Signals arriving in the primary motor cortex already contain the specific motor program, so all that is left for the primary motor cortex to do is to translate those into signals associated to individual muscles.

Let us imagine two extreme architectures:

1. **If any nerve detects a signal, all muscle cells contract.** This is the case with the nerve net in the previously discussed *Hydra* (see Chapter 5.1.1). As fewer motor neurons need to be activated, this option is extremely energy-efficient and quick: each movement could be done with maximum strength in the blink of an eye. Theoretically, you would need only a single nerve cell to control all muscles in your body. On the other hand, the only possible action you could take is to contract all muscles at the same time. Imagine you could fully exert your biceps with the same mental effort it takes to move your little finger. You would leave the gym with sore muscles but

---

[15]Rathelot and Strick, 2009.
[16]Rowe, Macrini, and Luo, 2011.

mentally, you would still feel refreshed. The downside is that you would not have any precision if all you could do is exert full force or exert no force at all.

2. **Each muscle cell is connected to an individual nerve cell coordinating whether or not it contracts.** A movement can involve any number of muscle cells in any combination and sequence, allowing very fine-grained control of the position of the limbs as well as the force exerted with each movement. This option is very energy intense, though. Not only would your body have to provide energy for your muscles to work, your nerve cells would take a similar amount of energy to operate the muscles. Our brain uses around 10 billion neuron cells to control 50 billion muscle cells. To reach a level of control where one neuron cell controls one muscle cell, our brain would have to be around 50% larger than its current size (our brain has around 86 billion neurons and would need around 126 billion neurons).

Our own primary motor cortex resembles the latter architecture more than the former. Compared to our more powerful cousins, the chimpanzees, our primary motor cortex is about 20% to 70% larger.[17] In chimpanzees, the relatively small number of motor neurons can be activated very quickly, allowing for greater strength. On the other hand, the higher number of motor neurons in humans allows for more precise movements and significantly less energy expenditure by the muscles. This gives us better control over how much force we want to exert—a crucial ability when it comes to using (or producing) more advanced tools like bows or spears.[18]

---

[17]Donahue et al., 2018.
[18]Walker, 2009.

5.2.2 The Rise of the Hominids

Further splits occurred about 15 million years ago (gibbons), 13 million years ago (orangutans), and 10 million years ago (gorillas), with the final split from chimpanzees and humans between four million and 13 million years ago. From there, we can trace back our human ancestors, with their most distinctive feature being a progressively increasing brain volume:

- *Australopithecus* (3.6 million years ago, 485 cm^3)

- *Homo habilis* (2.1 million to 1.5 million years ago, 650 cm^3)

- *Homo erectus* (2 million to 0.14 million years ago, 1100 cm^3)

- *Homo heidelbergensis* (0.7 million to 0.2 million years ago, 1230 cm^3)

- *Homo neanderthalensis* (250 thousand to 40 thousand years ago, 1500–1740 cm^3)

- *Homo sapiens* (today, 1425 cm^3)

With Neanderthals having had larger brains than humans, were they more intelligent than humans?

To answer this question, it is important to note that besides humans and their ancestors, other animals also have large brains. For example, bottlenose dolphins have a similar brain size as humans have. They have been observed helping out other dolphins as well as those of other species like humans (e.g., drowning divers), which points to them sharing our ability to understand what other beings are possibly thinking or experiencing.[19] But to compare brain sizes between humans and other animals, we need to include their *body size* into our calculation. Larger bodies produce a larger quantity of signals

---

[19]cf. White, 2007, p. 41.

that need to be processed and have more muscle cells that need to be activated. For example, while whales have a brain size more than five times that of humans, they are not necessarily quick (or creative) thinkers. They need that brain size to control the muscles in and the sensors of their huge body (around 50 tons). Thus, to get a sense about an animal's intelligence, we need to look at the total brain size divided by the body mass, and set it in relation to the average (for the species) brain size and body mass. This so-called *encephalization quotient* (EQ) provides such a mapping, making it possible to compare different animals' potential intelligence (see Figure 5.8).[20]

ENCEPHALIZATION QUOTIENT · The *encephalization quotient* (EQ) is a measure of relative brain size and is often used to convey how small or large a species' brain is compared to that of other species of similar body size.

Species	EQ	Cranial capacity
Human	7.4–7.8	1250–1450 cm^3
Neanderthal	7	1600 cm^3
Bottlenose dolphin	5.3	1350 cm^3
Homo erectus	5	1100 cm^3
Homo habilis	4.3	650 cm^3
Eurasian magpies	2.49	5 cm^3
Australopithecus	2.5	485 cm^3
Chimpanzee	2.2–2.5	330–430 cm^3
Gorilla	1.5–1.8	500 cm^3
Whale	1.8	2600–9000 cm^3
African elephant	1.3	4200 cm^3
Dog	1.2	64 cm^3
Cat	1.0	25 cm^3

Figure 5.8: Comparison of different species' brain size sorted by their relative encephalization quotient.

---

[20]Roth and Dicke, 2005.

As the table shows, when accounting for their larger body mass, the encephalization quotient of Neanderthals is similar to that of modern humans. And it has been shown that Neanderthals created and used tools like spears, possibly had language, and developed their own cultures. *So, how do we differ from Neanderthals?*

We find a clue by looking at gorillas. With an encephalization quotient of 1.5, they are at the lower end of the EQ spectrum—while clearly highly intelligent given that they can learn sign language and make and use simple tools. This points to the EQ not telling the entire story. Hence, scientists have divided the brain mass further into parts necessary for the maintenance and control of the body and senses, and those associated with improved cognitive capacities.[21] Subsequent studies on Neanderthal brains[22] have shown that they must have had significantly larger eyes (possibly to allow better sight for hunting where there was not much light) than those of humans. Also, the brain part responsible for image processing must have been larger, which left less brain mass for social relationship processing. The currently accepted theory is that humans used their social intelligence as an advantage over Neanderthals, ultimately replacing them—even though the latter possessed better sight and body control. Research seems to point to multiple migrations from Africa to Europe[23] and back from Europe to Africa[24] with interbreeding over a prolonged timespan.

Further research needs to be done in regard to arthropods (insects, spiders, etc.) and cephalopods (especially cuttlefish, squid, and octopodes). For example, jumping spiders are able to plan ahead and to apply hunting strategies, and even care for and nurse their young like mammals do.[25] This is reflected in their brain-to-body ratio: a jumping spider's brain requires so much space that it is distributed

---

[21]Roth and Dicke, 2012.

[22]Pearce, Stringer, and Dunbar, 2013.

[23]Kuhlwilm et al., 2016.

[24]L. Chen et al., 2020.

[25]Z. Chen et al., 2018.

throughout its tiny body. Another example would be octopodes; they have two-thirds of their nervous systems in their arms, resulting in a brain-to-body quotient comparable to that of humans. Octopodes can use tools, solve puzzles, recognize people, and plan ahead.

In the case of some animals, to counter the size limitations of the brain, the brain uses *cortical folding* to increase surface area and processing speed.[26] Imagine that very early during embryonic development, the neurons form a flat plane and then fold while the surface area is growing. This is comparable to creating towels by adding loops of thread (the folds) to a piece of cloth. They increase the surface area of the cloth without increasing the size of the cloth (the plane). Trees solve the problem of capturing sunlight in a way that is similar to the brain creating folds: to maximize the exposure of leaves to the sun, the tree creates branches which subsequently create branches and so on. This way, the distance from each leaf to the trunk is limited, while maximizing the amount of sunlight the tree can capture (see Figure 5.9). Objects like these are called fractals (see *Philosophy for Heroes: Continuum*).

Figure 5.9: This tree-like fractal covers a potentially infinite surface area while limiting the distance from the leaves to the trunk.

---

[26]Striedter, Srinivasan, and Monuki, 2015.

5.2.3 The Rise of the Humans

How could larger brains have helped early humans to survive? After all, we apply most of our intelligence to utilize complex language and tools. But those things were not available to early humans. And compared to other mammals, even compared to apes, humans are neither strong nor fast, we do not have body armor, we do not have claws or poison, we do not have wings or a strong sense of smell, and we cannot see well at night. At the same time, larger brains meant a higher energy need. In that regard, the question of how human intelligence evolved looks like another chicken-and-egg problem similar to the question of the origin of life (see *Philosophy for Heroes: Continuum*). We need to find an evolutionary path of incremental genetic changes, benefiting our ancestors at each step.

We have become a complex product of nature because our ancestors had to adapt and re-adapt again and again to ever-changing environments. Compare that to lifeforms that have not changed significantly for hundreds of millions of years: they have found a niche in which there was no evolutionary pressure to adapt to new challenges.

The important lesson is that evolution does not work in a directed way where all lifeforms become smarter over time. Brains are simply an adaptation to very specific problems. Each part of our brain addresses a particular need to process information in order to give us an edge over our predators and prey.

Hence, one approach to the question of how humans have diverged from other primates is to think about the influence of the *environment* on our ancestors' evolution. The split from the chimpanzees four million years ago probably happened between those of our ancestors staying in the forest and those going out into the savannah. As trees provided protection and sources of food, it is possible that the savannah was not their first choice. Perhaps a change in geology or climate caused the forest to recede. Later, population growth might have driven our ancestors out into the savannah; this is the *savannah hypothesis.*

> **SAVANNAH HYPOTHESIS** · The *savannah hypothesis* states that early humans evolved on the savannah and that many of the modern human's traits are a result of this adaption.

While the precise sequence of events is still debated—more recent evidence points to a much less abrupt transition from forest to the savannah[27]—the following traits were conducive for early humans to survive in the savannah:

Fire and cooking. Fire allowed early humans to be more active at night. It provided warmth and helped to fend off predators. With the ability to control fire, early humans were able to cook food. This reduced the time needed for digestion compared to eating raw foods, freeing up energy for the brain. While in other apes, the colon represents about 50% of gut volume, in humans, it is less than 20%.[28]

Bipedalism. Bipedalism probably developed before our human ancestors moved into the savannah. After all, apes are able to walk on two feet; they typically choose not to do so. Their muscle configuration makes walking less energy-efficient when compared to humans—just like we *could* walk on our hands, but it takes far too much energy.[29]

Endurance. Better muscle control due to the larger motor cortex (see Chapter 5.2.1) allowed our ancestors to travel longer distances —a very advantageous trait for living on the savannah.

---

[27]Domínguez-Rodrigo, 2014.
[28]Furness, Cottrell, and Bravo, 2015.
[29]Sockol, Raichlen, and Pontzer, 2007.

Cardiovascular system. We see additional optimization toward endurance in our heart which has adapted for moderate-intensity activity like walking, hunting, or farming. (By contrast, chimpanzees' hearts are optimized for short bursts of intense activity such as climbing and fighting).[30] This enabled our human ancestors to cover greater distances in the savannah, spreading their genes with tribes farther away. And while we cannot run faster than a cheetah in a sprint, we can outrun it in any hunt longer than a mile. Cheetahs are all about sneaking close to their prey and then sprinting toward it in one short burst.

Sight. In the forest, chimpanzees' main visual focus is spotting details. The global image of the forest as a whole is less important than identifying, for example, a snake, or fruits, or another chimpanzee hiding among the leaves. By contrast, the focus of humans adapted to a life in the savannah is a more comprehensive, global image. To coordinate a hunt, you need to keep your entire environment in mind and create a plan for how to approach prey from different directions. Studies have shown that humans are much better than other primates at integrating local visual information into a global whole.[31] Also, being on watch for predators or coordinating a hunt requires humans to constantly scan all directions. The anatomy of our eye sockets gives us a wider field of view compared to that of chimpanzees.[32] This is also more economical as we have to move only our eyes rather than our entire head.

Ranged weapons. It takes precision and perseverance rather than explosive strength when looking at an animal from a distance, feeling the wind, taking a wooden spear, calculating which of the dozens of muscles to stretch and release and in what sequence so that the stick hits its target, and then tracking the wounded animal over hours. In order to throw accurately, the brain has to calculate

---

[30]Shave et al., 2019.
[31]Denion et al., 2014.
[32]Imura and Tomonaga, 2013.

a sequence of nerve pulses to coordinate which of the hundreds of muscles should contract and in what order. To maximize the impact of a throw, the muscles have to work in harmony, just as performers in an orchestra play various instruments and create a harmonious sound. The comparison between throwing a spear and playing music is fitting as the same regions of the brain are used to produce both. While ranged attacks can be found in nature (tongues of chameleons, other primates throwing sticks and stones, archerfishes spitting water at bugs, pistol shrimps "shooting" air, jumping spiders jumping at their prey), humans are the *only* animal in nature to accurately throw things at long distances.

Language. Building a sentence requires advance planning and coordination. We need to use words in a specific order to carefully construct the sentence rather than trying to exert ourselves in one loud call. This uses the same mental machinery that is required to accurately throw. The *cognitive trade-off hypothesis* explains this difference between chimpanzees and humans by stating that our human ancestors exchanged (some of) their short-term memory for other abilities like abstract language and planning. Evidence for this is in (compared to chimpanzees) humans' significantly larger *angular gyrus* (see Chapter 5.4.1), which helps with tool use, reading, and language.[33] It is also home to our working memory (at least the phonological memory), which is superior in humans compared to chimpanzees.[34]

Reaction time and memory. While humans indeed require short-term memory for conversation, remembering telephone numbers, and reading, only a minimum amount is needed. Not only is a conversation—compared to the snap decision a chimpanzee has to make —stretched out over an extended period of time, but there is also always the possibility of asking questions.

---

[33] Fjell et al., 2013.
[34] Read, 2008.

Our ancestors in the forest faced a situation very different from those who went out in the savannah. In the forest, with trees blocking their line of sight, they had less time to react to sudden encounters with predators (or rivals). At the same time, they could use the trees to flee from their predators. This adaptation to quickly evaluate and react to a situation seems to be reflected both in today's forest-dwelling chimpanzees' strength as well as in their (compared to humans) superior short-term memory capacity. Experiments have shown that some chimpanzees need only 0.21 seconds (compared to humans needing at least 0.65 seconds) to remember the position and sequence of nine numbers on a computer screen.[35] This is similar to what a chimpanzee might encounter in the wild: imagine nine rival chimpanzees showing up. For the one chimpanzee defending itself, each second spent determining whether to flee to the trees or to prepare to fight might determine whether or not the chimpanzee loses territory, is injured, or even killed.

Games. Language also allows more educational *play*. While other animals (for example dogs or their puppies) have ways of signalling they want to play-fight, humans can develop much more complex games (e.g., hopscotch, fencing, martial arts, soccer, chess, etc.) to allow them to train their mental and physical abilities.[36] We can even use play productively: for example, we train people to operate in space while they are still safely on Earth.

Complex tool use. The mental machinery required to process language also allowed us to create complex tools. Sentences consist of subjects, objects, verbs, adjectives, and adverbs that connect with or modify each other. Similarly, tools consist of different objects that need to be connected. If you can imagine sentences that describe how entities interact with each other, you can imagine tools consisting of entities interacting with each other. Indeed, we can rely solely on language to describe the production of, for example,

---

[35]Inoue and Matsuzawa, 2007.
[36]Kerney et al., 2017.

a hand-axe or spear. While tool use is not something unique to humans, we have seen only a few examples in the animal world. For example, some monkeys have learned to dry certain nuts, and then later use specific stones to crack them open—a skill that can take years to learn properly.[37] But monkeys are probably not able to create something as complex as a bow and arrow (see Chapter 5.4.1).

Weapon evolution. Over time, our ancestors developed better and better weapons. For example, they added a sharp stone for additional weight, impact damage, and flight properties, and thus, invented the stone-tipped spear. Accurate ranged weaponry offered significant advantages to our ancestors. Not only did they become better at hunting, but also the prey's own defensive weaponry including hooves, claws, and fangs became useless against humans who were no longer within reach. Archaeological evidence puts spear use by our ancestors at as early as 500,000 BC.[38] This led to prey animals becoming more cautious. Individual animals that were more anxious had an evolutionary advantage over more courageous or curious animals. Thus, with each generation, they tried to stay farther away from anything that resembled a human. This put evolutionary pressure on humans to produce better and better tools, throw farther, and invent increasingly intricate hunting techniques and strategies. This created a predator-prey dependency. Like the evolution of the eye (see *Philosophy for Heroes: Continuum*), improvements in our ability to make and throw projectiles brought us an advantage at every step of the way.

Self-domestication. Chimpanzees are much more aggressive than humans, but they are also more hesitant to go into a fight, given that it comes with a significant risk of injury. Humans, on the other hand, evolved the ability to attack from a distance, attack together with others, and even to plan an attack in advance. This put anyone at risk, no matter his or her status. Anyone could be challenged as

---

[37] Luncz et al., 2017.
[38] Wilkins and Chazan, 2012.

long as there was support from other members of the tribe. This led to significant changes within tribes of humans: the most aggressive or anti-social members of the tribe could be singled out more easily. Just like we domesticated wolves by selecting the least aggressive pups from a litter, humans self-domesticated by removing the most aggressive members of their tribe. While (for the most part) peaceful *within* the tribe, humans became efficient hunters for everything outside of their tribe. We still have to grapple with this dual nature of humanity—sharing a strong sense of community, while also expressing an "us versus them" mentality. On the one hand, we can easily make peace with the people around us; on the other hand, we can *rationalize* (by mentally degrading them) killing animals and killing humans from other tribes. While we might look down at chimpanzees for their in-the-moment aggression and impulsivity and see ourselves as the pinnacle of evolution, we also have to remember how easily humans can suppress their empathy once they have judged someone as "sub-human."

Interestingly, studies have shown that in mammals, certain genes regulate both aggressivity and aspects of anatomical features of the face. This genetic connection explains why we have shorter faces and smaller teeth compared to our primate cousins. Similarly, many domesticated dogs look less threatening than wolves. If our ancestors did not specifically breed wolves, it is conceivable that the least aggressive (and thus also least aggressive-looking) wolves were able to approach a human camp without being attacked by humans. A connection between behavior and facial structures can also be seen in humans. People who are missing a copy of the BAZ1B gene have the *Williams-Beuren syndrome* and are more talkative, outgoing, and less aggressive. They also have rounder faces with shorter noses, full cheeks, and wide mouths with full lips. Vice versa, people with additional copies of the BAZ1B gene (the *7q11.23 duplication syndrome*) tend to be aggressive, have difficulties socializing, and their facial features are also affected.[39]

---

[39]Zanella et al., 2019.

Rules and laws. Over time, human civilization has replaced the lethal way we deal with aggressive members of our tribes with methods of coping with our emotions: culture, customs, rules, and ultimately, laws. Those of our ancestors who were able to reflect upon their own status and position within society, and how it might feel to be another member of society, ultimately prevailed. As the anthropologist Richard Wrangham put it, "Those who followed the rules were favored by evolution."[40]

Compared to our primate cousins, humans are actually *much less* eager to change the pattern of how they carry out a task.[41] In a study, the participants—humans, rhesus macaques, and capuchin monkeys—had to select three icons in sequence to score a point (or get a banana) in a set of 96 trials. The third symbol showed up once the first two were selected in the correct order. In the second set of the trials, the third symbol showed up immediately; participants could either select all three symbols in order or select only the third symbol to get the reward directly. Most monkeys switched to the more efficient strategy of selecting the third symbol without delay, while humans tended to stay with their previous strategy. Preference for the familiar over a new approach was also confirmed in a second study comparing humans with chimpanzees.[42]

Protected childhood. Compared to babies of other mammals, human babies are totally dependent on their parents. This becomes obvious when comparing human babies to chimpanzee babies: without their mother, human babies are helpless, while chimpanzee babies are at least able to follow or hold onto their mother from early on. Intelligence-wise, human children catch up with chimpanzee children only at the age of one or two years. This is because the brain of a human baby continues to proliferate after birth; by contrast, in terms of growth, chimpanzee brains level off very soon after

---

[40]Wrangham and Grolle, 2019, cf.

[41]Watzek, Pope, and Brosnan, 2019.

[42]Pope et al., 2020.

birth. We should not look at prolonged childhood being an evolutionary mishap or obstacle, though. Instead, it is more a testament to the success of human parents to provide the necessary protection and nutrition during the earliest periods of their children's lives. It is an expression of the long-term *investment* into the brain development of the child.

Giving a baby the safe space to develop his or her brain without the need to survive independently gives humans an evolutionary edge. It is easier for the brain to put down new neural pathways in a part of the brain that is not being used at the time, just like it is easier for a construction company to replace rail tracks when no trains are using them. We have to keep in mind that we cannot just shut down our brain for a few days for architectural changes. While some cleanup processes happen during our sleep, we still have to be ready to respond within seconds when being awakened by, for example, the sound of a possibly dangerous animal or another human. The requirement to care for our children seems to be another evolutionary factor driving our longevity. As grandparents, we can spend resources on caring for our grandchildren or other relatives. This is not exclusive to humans; it has been shown that among killer whales, post-menopausal whales provided significant survival benefits for their grand-offspring.[43]

With our evolutionary history in mind, let us now take a closer look at the most complex part of our brain, the *neocortex*. In Chapter 5.3, we will look at how the brain processes sense data, while in Chapter 5.4, we will cover how the brain initiates and executes movements. Taking everything together, we will examine how the brain can recognize itself in the mirror in Chapter 5.5. Once we understand the brain parts involved in the input of sense data and output of motor actions, we can then proceed in Chapter 6 with the central topic of this book, consciousness.

---

[43]Nattrass et al., 2019.

5.3 A Glimpse into the Brain

How does the brain perceive its environment?

To grasp the brain's underlying functionality, it is best to look at how information flows through individual parts of the brain. While we have already discussed some aspects of the visual cortex, let us revisit it in more detail to see how different systems in the brain interact.

LOBE · A *lobe* is an anatomical division or extension of an organ.

OCCIPITAL LOBE · The *occipital lobe* is part of the *neocortex* and contains the *visual cortex* which is responsible for processing visual sense data.

CEREBRAL HEMISPHERES · The *cerebral hemispheres* consist of the *occipital lobe*, the *temporal lobe*, the *parietal lobe*, and the *frontal lobe*. The two hemispheres are joined by the *corpus callosum*.

Figure 5.10 shows the architecture of the visual system.

- **Right visual field:** Light from the right visual field hits the left sides of the retinas of each eye.

- **Left visual field:** Light from the left visual field hits the right sides of the retinas of each eye.

- **Left retina sides:** Sense data from both left retina sides are communicated through the optic chiasm and combined in the *visual cortex* of the left cerebral hemisphere of the *occipital lobe*.

- **Right retina sides:** Sense data from both right retina sides are combined in the visual cortex of the right cerebral hemisphere of the occipital lobe.

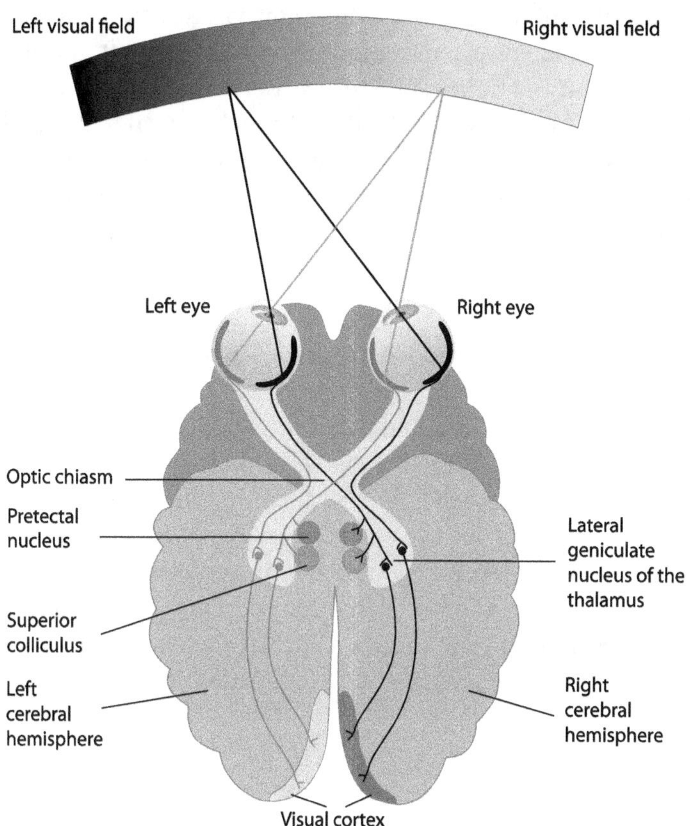

Figure 5.10: The basic visual projection pathway from our eyes to our neocortex. The eyes' lenses project light onto the retinas of each eye. There, light-sensitive cells translate light into electrical impulses. The left side of the projected image is sent to the right hemisphere, and the right side of the projected image is sent to the left hemisphere (image source: Shutterstock).

Ultimately, information from the right visual field is processed in the left visual cortex, and information from the left visual field is processed in the right visual cortex. Between the optic chiasm and the visual cortex, the signal passes through the left and right lateral geniculate nucleus. There, the sense data is pre-processed and transferred to various brain parts, with the visual cortex the most important one. Pre-processing means that the brain analyzes the images and modifies them for further processing by other brain parts.

There are a number of possible reasons evolution led to this rather convoluted architecture where the optic nerves of both eyes cross in the optic chiasm and split the information from the eyes' left and right visual fields. If the left eye were directly connected with the left visual cortex and the right eye directly connected to the right visual cortex, losing one eye would mean that one visual cortex would either no longer receive any input, or it would receive the input too late because it first had to be transferred from the other hemisphere. The crossing in the optic chiasm enables the visual system to work even in the case when only one eye functions properly.

5.3.1 Color Perception

To understand vision, we first need to understand color and light. Light rays are actually electromagnetic waves. Depending on the length of the waves, we experience them as different colors on the color spectrum (so-called spectral colors, see Figure 5.11). To perceive colors, our eyes have photoreceptor cells sensitive to red, green, and blue light (see Figure 5.12). In addition, shorter wavelengths activate both the red and blue photoreceptor cells, making them look purple.

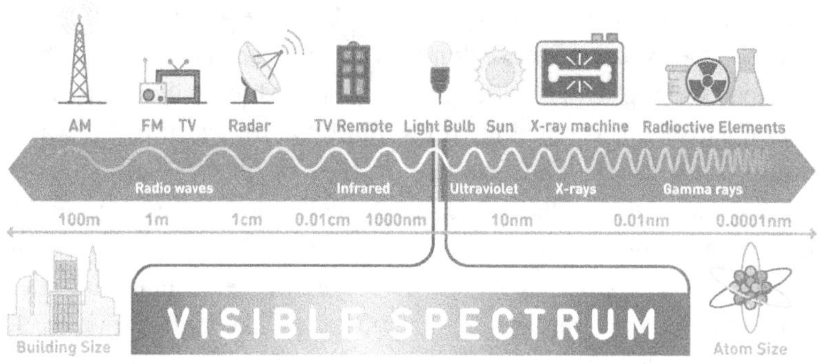

Figure 5.11: The electromagnetic spectrum (image source: Shutterstock).

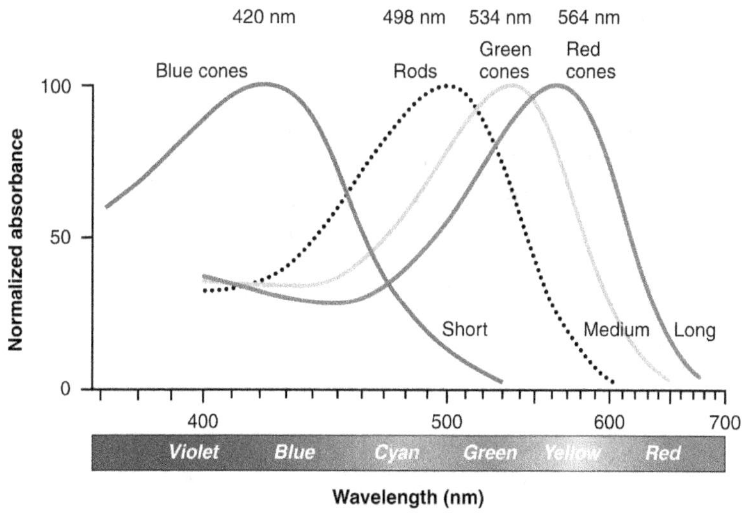

Figure 5.12: Sensitivity of the cones and rods differs depending on the wavelength (image source: Openstax, Rice University).

Now, imagine throwing a stone into a lake. Depending on the size of the stone, water waves of different size will emerge. A large stone will lead to long water wavelengths, a small stone will lead to short water wavelengths. If you threw multiple stones into the water, there would still be water waves, but they would overlap with each other. Just like no single stone can create such an overlapping of water waves, no single light source can create all the colors we can perceive. This is why beyond the colors on the electromagnetic spectrum, we also experience *extra-spectral colors* like white, gray, black, pink, or brown when different photoreceptor cells are activated in combination (see Figure 5.13). Those colors could be compared with multiple stones thrown into the water. For example, a pink flower reflects red light waves, but also reflects green and blue light waves. Similarly, white, gray, and black are the product of different wavelengths at the same intensity, and brown is simply orange at low light intensity.

Red	**Green**	**Blue**	**Interpretation**
100%	0%	0%	red
0%	100%	0%	green
0%	0%	100%	blue
100%	100%	100%	white
50%	50%	50%	gray
0%	0%	0%	black
100%	75%	80%	pink
100%	60%	0%	orange
75%	20%	0%	brown

Figure 5.13: A translation of the activated photoreceptor cells to the subjective experience of color.

That extra-spectral colors are a combination of different wavelengths of light was not discovered until Newton's famous prism experiment in 1666. Before that, it was thought that prisms *produce* colors, but it was Isaac Newton who showed that a prism merely *splits* light into its spectrum. By adding a second prism, Newton proved that the red light from the first prism produced only red light in the second prism, and that he could recombine different colors of light back into white light (see Figure 5.14). This experiment became a symbol for the Scientific Revolution because it replaced a subjective understanding of light with objective, observable facts. The existence of extra-spectral colors show that our subjective experience of the world is pre-processed. The LGN integrates the information from the retina into color information. For example, if we look at a purple van, our red and blue photoreceptor cells are activated. But no matter how close we get to the van, we never see separate blue or red color elements. What we see is pre-processed for us through the combination of different sources of sensory information into new data, the color purple.

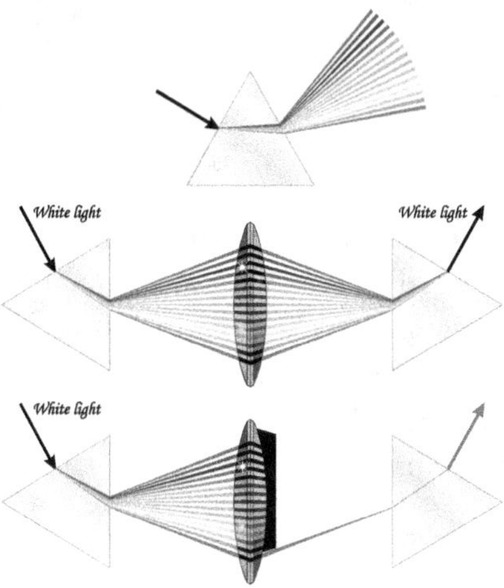

Figure 5.14: Newton's experiment that demonstrates how white light actually consists of light of different colors (image source: Shutterstock).

More complex processing involves the inclusion of the three-dimensional data the LGN has derived from sense data. It is used to further modify the way the LGN integrates color perception. Consider the checkers shadow illusion in Figure 5.15: while in the first picture it looks like a checkerboard with a three-dimensional black ball throwing its shade over the board, the second picture shows how both marked squares were printed (or are displayed) with the identical shade of grey. Our brain processes the image to give us the impression of how the checkerboard would look without the shadow. Without this processing, it would look like the shadow was actually printed on the checkerboard.

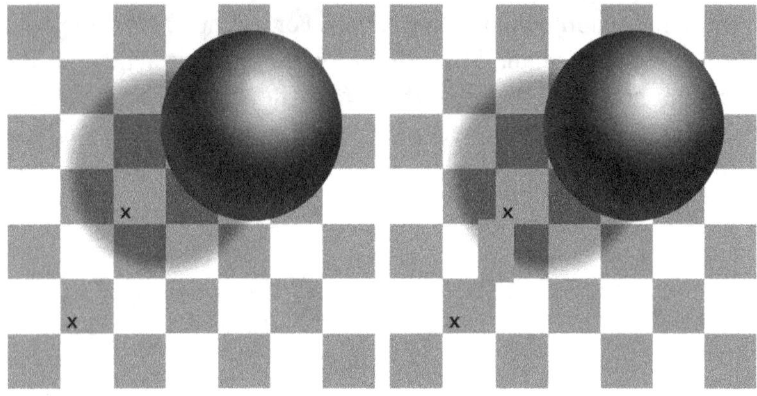

Figure 5.15: The checkers shadow illusion misleads the viewer into thinking that the two-dimensional image of a three-dimensional checkerboard is evenly colored black and white. The second image shows that the squares inside and outside the shadow are identical in color (image source: Shutterstock).

5.3.2 The Brain as a Prediction Machine

Processing visual and auditory data requires time. Studies have shown that the brain needs around 190ms for visual stimuli and 160ms for auditory stimuli.[44] This poses a significant problem: any decision the brain makes is calculated based on old information. A delay of 190ms does not sound like very much. But imagine a ball that is thrown at you with a speed of 100km per hour (or about 15 meters per second). In 190ms, the ball travels a distance of around five meters. With that delay, you would always catch the ball too late. This problem becomes even more complicated if you are moving. To deal with this delay, the brain tries to predict the positions of where objects will be in the near future based on their current speed. This helps other parts of the brain to make more accurate decisions, for example, catching the ball in the right moment.

There are a few instances where this prediction process of the brain fails noticeably. Consider the *Hering illusion* in Figure 5.16: the straight lines near the central point appear to curve outward. Our visual system tries to predict the way the underlying scene would look in the next instant if we were moving toward its center. The cost of predicting the future to reduce reaction times is that we sometimes experience this correction as an optical illusion.[45]

While we can establish that optical illusions are the product of pre-processing and ultimately useful for us, the question is *why* there is a difference between how we *experience* what we see and what is actually there. One could argue that it is just an optimization or filter by the brain for specialized applications (movement, 3D, faces, and so on). But this does not explain why this optimization feels *real* to us, even when we have evidence to the contrary.

---

[44]Welford, 1980.
[45]Changizi et al., 2008.

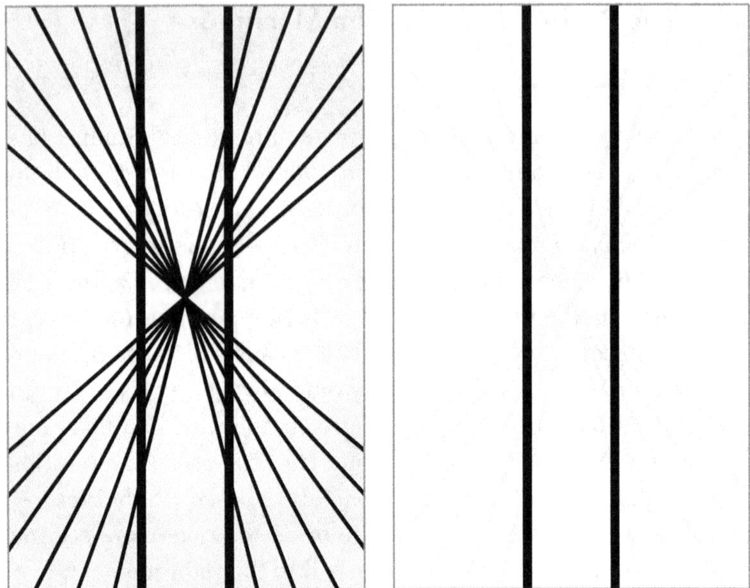

Figure 5.16: Two parallel lines look skewed when the mind is given a spatial context (image source: Shutterstock).

5.3.3 The Ventral and Dorsal Streams

The output from the visual cortex continues as two separate data streams:

- The ventral (lower) stream (the *what*) through the temporal lobe (see Figure 5.17).; and

- The dorsal (upper) stream (the *where* and *how*) through the parietal lobe.

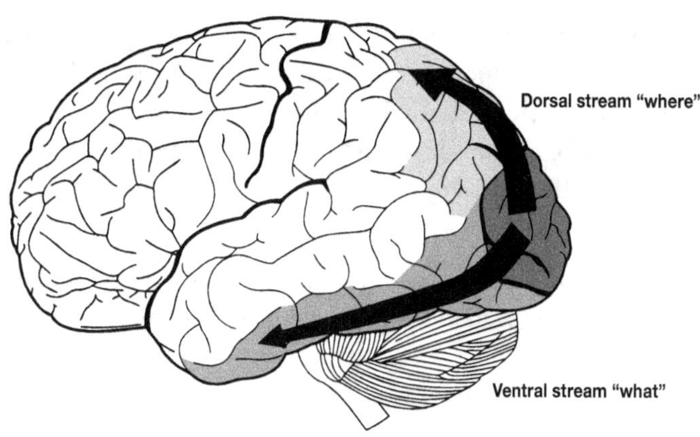

Figure 5.17: Image showing the ventral stream (the *what*, at the bottom through the temporal lobe) and the dorsal stream (the *where* and *how*, at the top through the parietal lobe) in the human brain visual system (image source: Selket, Wikimedia).

The *temporal lobe* processes visual information to categorize *what* things are, while the *parietal lobe* integrates the spatial information into a map of *where* and *how* things are. For example, looking at an apple on a table, the temporal lobe identifies that there is an apple and a table, and the parietal lobe identifies that the apple is on top of the table.

TEMPORAL LOBE · The *temporal lobe* is the part of the neocortex that deals with the "what": long-term memory, and object, face, and speech recognition.

PARIETAL LOBE · The *parietal lobe* is the part of the neocortex that deals with the "where," especially the location of entities, the "how," as well as touch perception.

Compared to the dorsal stream, the ventral stream is processed more quickly. It is more important to know what you see (a tiger, a fire, an acquaintance, etc.) than where it is. This idea of prioritization of the processing of the sense data can also be found in the general architecture of the visual system.

One could ask why our visual cortex is at the very back of our brain given that it takes valuable time for a signal to pass through the brain. Why not have all visual processing at the front or at least directly behind the eyes? Given the actual architecture, processing in the visual cortex seems to have a low priority. It is furthest away from the eyes—the opposite of what we would expect from an organ that can require an immediate response.

Looking at it from an engineering point of view and turning the question around, we would ask: what essential functions of the visual system should be put at the front? That is, reflexes to close your eyelids to protect your eyes, to combine information from your left and right eye, to turn your head to a source of movement or sound, to focus your lenses, and to constrict your pupils to protect your retina. All brain parts responsible for these abilities are positioned around the superior colliculi near the eyes. They provide us with reflexes and mechanisms to protect the eyes and refocus them.

After the initial processing in the superior colliculi, the processing goes through the LGN. There, a three-dimensional representation of the world is created, with just enough detail to allow for quick —possibly life-saving—reactions. For example, in a ball game, if we always had to first conceptualize that a ball is flying at us and plan for its arrival, our reactions would be very slow. Learning to catch a ball requires us to bypass conceptualization and just "do"—trusting our instincts supported by early calculations in the LGN before the information even reaches our visual cortex.

5.3.4 The Temporal Lobes

The temporal lobes of the cerebrum are positioned on the left and right side of the brain, near the ears. They process auditory signals and are responsible for identifying *what* is being said (or seen, as part of the ventral stream). Not only does this brain part identify what you see, but it also maps it to language (in *Wernicke's area* of the left temporal lobe). If Wernicke's area is damaged, you would use *individual* words correctly, but in combination, the words may not make any sense.

> **WERNICKE'S AREA** · *Wernicke's area* is located in the left side of the temporal lobe. Its function is the *comprehension* of speech. Damage to Wernicke's area leads to people losing the ability to form meaningful sentences. It is connected to Broca's area, which is responsible for muscle activation to *produce* speech.

The brain part in the right hemisphere corresponding (homologous) to Wernicke's area deals with subordinate meanings of ambiguous words (for example, "bank" refers to a financial institution but could also refer to a river bank).[46] With the ventral stream, you can identify that you are seeing a tree and connect it to the abstract concept of a tree. It also allows you to map the concept of a tree to the image of a tree or to the sound of the word "tree." As we have learned in *Philosophy for Heroes: Knowledge*, this mapping is ultimately connected to a past experience. In terms of learning languages, we connect a word with the experiences we had when hearing or reading the word in the past. Someone pointed to a tree, said "tree," and we connected sound and image with the concept of a tree. As this suggests, the temporal lobe is essential to processing memories. This is supported by the fact that the temporal lobe is also connected to the hippocampus, providing spatial memory and short-term memory, as well as helping to create long-term memories.

---

[46]Harpaz, Levkovitz, and Lavidor, 2009.

Figure 5.18: A basic drawing of a kitchen. Despite most of the visual information (textures, colors) being stripped from the image, we can immediately classify it correctly (image source: Shutterstock).

One could argue that a comic strip is the brain's internal representation of what is left after the brain has processed an image—just like the word "tree" is a representation of a real tree. To conceptualize the environment, the brain tries to strip all superfluous information from an image.[47] For example, Figure 5.18 shows a simplified drawing of a kitchen. We can immediately recognize it as a kitchen. Abstract symbols that are stripped of all superfluous information (colors, textures, etc.) are even *easier* (especially with less ambiguity) to recognize, hence their use in street signs.

---

[47] Morgan, Petro, and Muckli, 2019.

5.3.5 Facial Recognition

Given that we can remember thousands of faces despite them differing only minimally, it is no surprise that we have specialized mental machinery specifically for faces or face-like structures. The temporal lobe contains the *fusiform face area* that deals only with identifying faces. People who lack this ability of the brain to pre-process faces have prosopagnosia ("face-blindness"). Imagine that everyone you meet is wearing a mask: you would have to remember what clothes someone usually wore, how her voice sounded, and categorize people by their hairstyle, height, or body type. Similarly, if we have not encountered enough people from a particular background (Asian, African, European) to have learned to distinguish her facial features, we might have difficulty recognizing individual differences.

Our face recognition is so important that it goes as far as seeing faces where there are none. For example, a standard American power outlet is just that, a power outlet (see Figure 5.19). We are "projecting" that it is a surprised face, although we are absolutely sure that there is certainly not a (human) face in the wall. It seems that we share this ability to recognize basic facial features (two circles and a mouth) at least with reptiles, going back more than 300 million years in our evolutionary history.[48] Looking for faces everywhere can help us to quickly recognize a friend (or enemy)—at the low cost of identifying faces when there are none.

---

[48]Versace, Damini, and Stancher, 2020.

Figure 5.19: A standard American power outlet, which looks like a surprised face (image source: Shutterstock).

The downside of this pre-processing is that we can have a harder time focusing on details of a (known) face. The "Thatcher effect" is a demonstration of this. Looking at Figure 5.20, you can see a young woman's face with a neutral expression. But turn this book upside down, and you will see a woman with a creepy grimace. This is because your ability to recognize faces is optimized to recognize people standing on their feet rather than hanging upside down from a tree. For the neural networks our brain uses, it would take extra effort to check whether or not mouths and eyes are oriented in the right direction. People with prosopagnosia are affected by the Thatcher effect, too, as their only struggle is with the classification of the face as a whole. They still use the same machinery as people without prosopagnosia to classify individual facial features (like the eyes or mouth). It is like the opposite of not being able to see the forest for the trees: we get the meaning of something ("this is a face of a young woman") but miss the details ("her eyes are upside down").

The *extrastriate body area* and the *fusiform body area* are similar to the fusiform face area. They are located in the visual cortex near the fusiform face area and deal with recognizing body parts and body shapes, and analyzing the relationship of moving limbs. Weaker connection between the extrastriate body area and the fusiform body area can lead to misjudgements of one's own body size and to illnesses like anorexia nervosa.[49]

---

[49]Suchan et al., 2013.

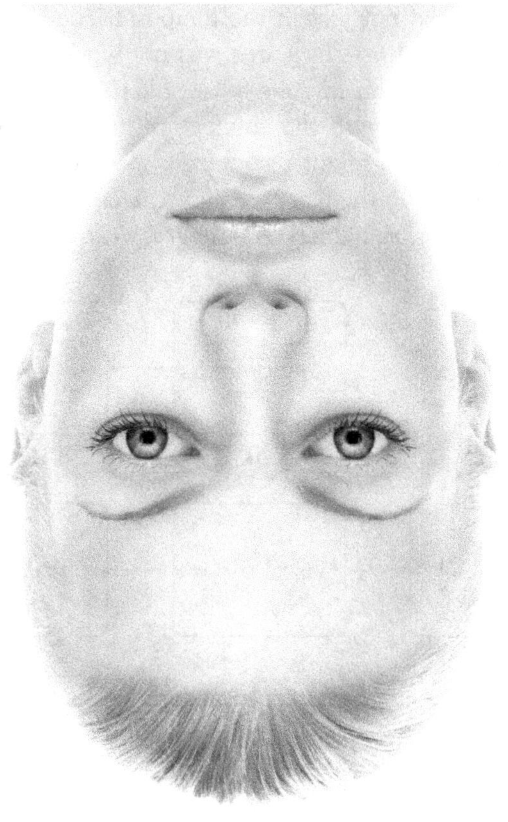

Figure 5.20: Turn your book upside down to experience the "Thatcher Effect." It demonstrates how our visual system is optimized to recognize people standing on their feet rather than hanging upside down (image source: Shutterstock).

5.3.6 Delusions

Faces are evaluated not just in the temporal lobe but also in the amygdala. If the fusiform face area was not working properly, we would still experience an emotion but would not recognize the face. If there is a problem with the connection between the thalamus and the amygdala, we might recognize a face but would lack the emotional response to the face (see Figure 5.21). This can lead to a *monothematic delusion*, namely the *Capgras delusion*.[50]

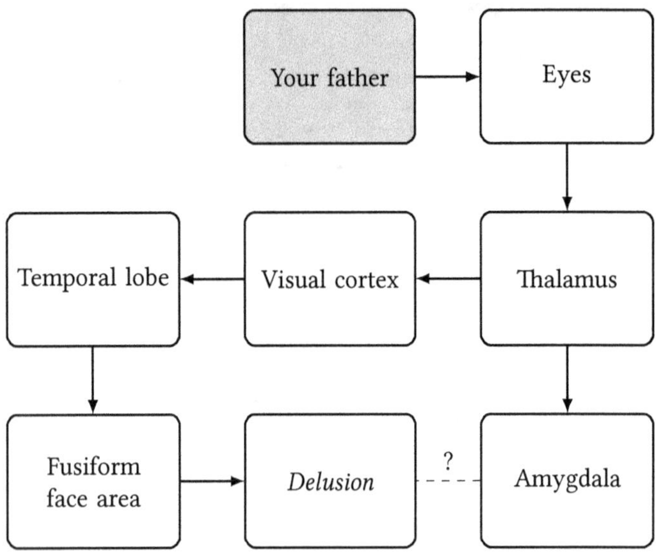

Figure 5.21: If your amygdala does not provide the right emotional information related to a perception, the conflict with the information from the fusiform face area can lead to experiencing the Capgras delusion (in which you assume that the person is an imposter).

---

[50]Ramachandran, 1998.

> **Monothematic delusion** · A *monothematic delusion* is a delusion focused on a single topic. A delusion is a firm belief that cannot be swayed by rational arguments. It is distinct from false beliefs that are based on false or incomplete information, erroneous logical conclusions, or perceptual problems.

> **Capgras delusion** · A person suffering from *Capgras delusion* can recognize people who are close to him, but he thinks they have been replaced by clones or doppelgangers. One cause for this condition is a damaged or missing connection from the amygdala that leads to a person having no emotional connection to those whom he sees.

The Capgras delusion is the belief that a person emotionally close to you (for example, your father) has been replaced by an imposter. You definitely see that the person is exactly as you have him in your memory; it is just that you no longer feel an emotional connection to him. For the brain, the most apparent (although weird) solution to solve this conflict is to assume that someone is impersonating your loved one. You cannot give clear reasoning for it, but based on the facts (the sense data identifying the person plus the lack of an emotional connection), it is the most logical conclusion. This can also happen in healthy people with movie actors. Having built an emotional connection to the character a person plays, there is a disconnect when meeting the actor in person: the actor looks exactly like the character but the emotional connection to the actor is missing. The best explanation the brain might come up with is that the person whom you are seeing (and who is the actor) is an imposter. This is more probable when meeting the actor outside the usual environment (convention, conferences, movie award events, etc.) in daily life (at the grocery store).

Other noteworthy delusions are:

- Fregoli delusion (all the people you meet are actually the same person in disguise);

- Syndrome of subjective doubles (there is a doppelganger of yourself acting in your name);

- Cotard delusion (you are dead or do not exist);

- Mirrored-self misidentification (the person in the mirror is someone else); and

- Reduplicative paramnesia (a place or object has been duplicated, like the belief that the hospital to which a person was admitted is a replica of an actual hospital somewhere else).

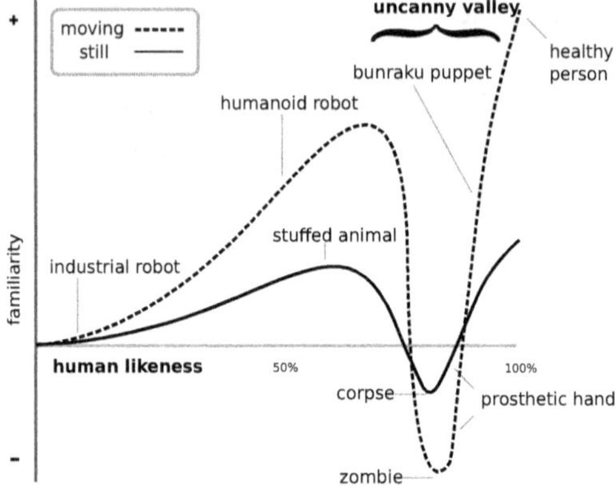

Figure 5.22: The "uncanny valley" effect with human-like dolls or robots (image source: Karl MacDorman).

Somewhat related to delusions is the "uncanny valley effect."[51] We have no problem dealing with dolls. However, when they become too human-like (but not fully human-like), we experience feelings of eeriness and revulsion (see Figure 5.22). This is why zombies (and to an extent, human-like robots) are used in horror movies. Even modern computer graphics artists have serious problems creating believable faces. We can notice the smallest deviation from reality even if the face geometry fully matches the human face. The artist has

---

[51] Rosenthal-von der Putten et al., 2019.

to hit all the marks when it comes to lighting, mouth movements, nose shadows, skin pore structure, and so on. This is of no surprise as research points to face recognition being highly evolved as it is used also to detect kinship[52] which is an evolutionary advantage in taking care of relatives as well as for mate selection.

> **UNCANNY VALLEY EFFECT** · The *uncanny valley effect* refers to the negative reaction to dolls or robots that are very (but not fully) human-like. Possible reasons for this reaction are an inbuilt instinct to avoid corpses, and the inner conflict and discomfort of switching back and forth between seeing a being as fully human or a lifeless object.

The uncanny valley effect is similar to the Capgras delusion in regard to there being two pieces of conflicting information in the brain. One pathway tells you "This is a human!" while another warns "No, that is no human," and the brain has to sort it out somehow, leaving you with a strange feeling of uncertainty. Some of us might have experienced the sudden shock at night when we see a person standing in our living room, only to discover that it is just the clothes rack. The physical presence of a person implies relationships, conflicts, alliances, and enemies. Humans can be more dangerous than the wildest animal, hence anything that resembles a human gets the most immediate attention. In addition, some of the discomfort could also stem from pathogen avoidance (a nearly human-like robot could also be seen as a real human with a serious illness or even as a corpse).[53]

With perception on the one side, we now need to look at consciousness from the other side: action. To act in this world, we need to know that we have a body and know how to use it. We need to differentiate between self and other by creating a so-called "body schema" which we will discuss in Chapter 5.4.

---

[52]Kaminski et al., 2009.
[53]Moosa and Ud-Dean, 2010.

5.4 Creating a Body Schema

How does the brain know what part of the body it belongs to?

To use our body, our brain needs to build a *body schema*. Without such a schema, we would experience our body only in an abstract sense, the same way we know we have a pancreas or liver.[54] Having a clear idea what part of the body is actually part of oneself and where each part of the body is located and how it works is essential for survival. For example, consider passing through a doorway. Damage to the brain parts responsible for creating the body schema can lead to anorexia nervosa. In someone suffering from this condition, the brain miscalculates the size of the body and the person thus may hesitate to walk through doorways thinking he will not fit.[55]

> **BODY SCHEMA** · The *body schema* is the brain's simplified description of the status of the body. It is built by correlating what we see and feel our body is doing with signals that the brain sends to the muscles.

Imagine seeing an apple. In order to grasp it, your brain needs to build a body schema using these six pieces of information:

- You have a body;
- The position of your body (and limbs);
- There is an apple;
- The position of the apple;
- The position of the apple relative to your body; and
- The position of your arm relative to your body.

---

[54] Graziano, 2017.
[55] Guardia et al., 2012.

To gather this information, you need both the temporal lobe ("What do I see?" to identify the object as an apple), and the parietal lobe ("Where do I see it?" to gauge the position of the apple). Without this body schema, you would have to move your arm and then keep readjusting the muscles so that the image of your hand gets into a position to grasp the apple. We have to rely on this clumsy method when, for example, manipulating something which we only see in a mirror. We have to consciously flip the image (and translate left to right and right to left) instead of relying on our body schema.

5.4.1 The Parietal Lobe

The parietal lobe of the cerebrum is positioned at the top of the brain. It is responsible for processing information about the "*where*" and thus helps us to recognize our limbs. Any damage to the *posterior parietal cortex* (the area at the end of the parietal lobe directly adjacent to the occipital lobe) can cause disorders like simultanagnosia (the inability to recognize the forest for the trees), akinetopsia (the inability to perceive motion), apraxia (the inability to produce volitional movement), or optic ataxia (the inability to use visuospatial information to guide movement). These findings support the idea that the parietal lobe is responsible for further processing of visual information from the occipital lobe, especially regarding positioning and movement. At the other end of the parietal lobe is the *postcentral gyrus* (adjacent to the frontal lobe). It contains the *primary somatosensory cortex* which (like the visual cortex for visual information) is the main sensory receptive area for the sense of touch. Different parts of the primary somatosensory cortex are specialized for sense input from specific body parts. A disproportionately large number of neurons are allocated to process sense data from the lips, tongue, face, and hands.

PRIMARY SOMATOSENSORY CORTEX · The *primary somatosensory cortex* in the parietal lobe deals with the processing of tactile sense input. Each body part is represented in the primary somatosensory cortex. The size of each representation correlates with the sensitivity of the body part to tactile stimulation (for example, lips and hands have a larger representation than other body parts).

Located between the primary somatosensory cortex and the occipital lobe, the parietal lobe contains the *superior parietal lobe*. It is involved in creating and maintaining a mental representation of the internal state of the body. Given its location and role in the perception of self, the superior parietal lobe is also implicated in episodic memory, visuospatial processing, and self-reflection.

SUPERIOR PARIETAL LOBE · The *superior parietal lobe* deals with creating and maintaining a mental representation of the internal state of the body.

The parietal lobe also contains a brain part called the *angular gyrus* which receives visual, auditory, and somatosensory information from adjacent brain areas. It deals with putting things into relationship with each other:

- The *left angular gyrus* is responsible for transferring visual information to Wernicke's area. It is involved in combining letters into words and words into sentences. It also deals with recognizing the positioning of numbers relative to each other, which is important for mathematical calculations.

- Research has shown that direct electrical stimulation of the *right angular gyrus* can produce a temporary out-of-body experience.[56] This points to the right angular gyrus being involved in the relationship of the self to the external world.

---

[56]Blanke et al., 2002.

> **ANGULAR GYRUS** · The *angular gyrus* combines visual, auditory, and somatosensory information and puts things into relationship with each other. The *left angular gyrus* deals with relationships in the external world, especially relating to words and letters, and the *right angular gyrus* deals with the relationship between the self and the external world.

The *supramarginal gyrus* (SMG) is adjacent to the primary somatosensory cortex (sense of touch) and the angular gyrus (limb positioning). Its main tasks are to determine and interpret the status of your and other people's limbs. The left and right SMG have different specializations:

- The *left SMG* is particularly involved in tool use. This is because tools require a specific grip and can act as an extension of your arms and hands.[57]

- The *right SMG* interprets the postures and gestures of other people. If it is disrupted (for example, through magnetic stimulation during an experiment), subjects begin to put more weight on their own point of view. They lose the ability to assess another person's feelings by looking at that person's face, body posture, and gestures. Being blind to other people's emotional states, subjects start to project their own emotions onto others.[58] Accordingly, training our right SMG through mindfulness exercises like certain forms of meditation not only increases its size,[59] but also seems to improve our ability to read other people's emotional states. A limitation of this study is that (short) mindfulness exercises seem to have the opposite effect on people with a narcissistic personality disorder, enabling them to focus *even more* on themselves.[60]

---

[57] Andres et al., 2017.
[58] Silani et al., 2013.
[59] Sevinc et al., 2019.
[60] Ridderinkhof et al., 2017.

> **SUPRAMARGINAL GYRUS** · The *supramarginal gyrus* (SMG) creates an internal representation of the *limbs* of the body, similar to the adjacent superior parietal lobe (which creates a representation of the *internal state* of the body). It supports tool use (left SMG) and interpreting the emotional state of other people based on their postures and gestures (right SMG).

Interestingly, both the left and right SMG also seem to be relevant for recognizing and reading words.[61] In addition, at least the left SMG seems to be involved in the phonological memory which we will discuss in Chapter 6.3.2.[62] Given that the angular gyrus and Wernicke's area are adjacent to the supramarginal gyrus, we can see the evolutionary connection we have mentioned in Chapter 5.2.3 between gestures, posture, tool use, and language. The same mental machinery that builds and uses the rules for how limbs connect and move relative to each other is re-used for building sentences and tools.

For the brain, at a conceptual level, connecting limb–joint–limb is the same as connecting subject–verb–object (grammar). Similarly, mechanics could be seen as a form of grammar for how different parts of a machine or tool interact. For example, to build and handle a bow, it requires you to imagine at least five different interacting parts (arc, bowstring, projectile, feathers, arrow-tip). This depends on the mental ability to form concepts and simulate how those concepts could work together—parallels to language and grammar are obvious. The angular gyrus seems to also be involved in the production (and even the imagination) of music—again, activities that require the application of grammar-like rules.[63]

Given its reliance on grammar rules, language is highly predictive and we can process it efficiently. Often, we already know what words come next before others have finished a sentence—especially

---

[61] Hartwigsen et al., 2010; Stoeckel et al., 2009.
[62] Deschamps, Baum, and Gracco, 2014.
[63] Tanaka and Kirino, 2019.

when also taking into account our ability to anticipate what other people might be thinking. This comes at the risk of being misled by so-called "garden path sentences." Try reading the following sentences without getting confused (meaning your mind predicts something and you start stumbling over the words at the end):

- "The old man the boat."
- "The girl told the story cried."
- "After the student moved the chair broke."
- "Fat people eat accumulates."
- "The complex houses married and single soldiers and their families."
- "The horse raced past the barn fell."

For example, in the first sentence, "The old man the boat," we first assume that "old" is an adjective describing "man" and then predict a verb will follow to describe what the old man does with the boat. Only after carefully reflecting on the structure of the sentence we might notice that "the old" refers to "old people," while "man" is the verb of the sentence. Similarly, in the other sentences, there is confusion between active and passive ("told"), subject and object ("chair"), adjectives and nouns ("fat", "complex"), and subordinate clause and main clause ("raced" and "fell").

5.4.2 The Premotor Cortex

While we know now where things are and how they are positioned relative to each other, the question is still open which of those things are our own limbs. Just because we see an arm does not automatically mean it is our arm. If you think that this is obvious and that there is no need to think about which limbs are yours, it is exactly because you are using your body schema to determine that.

A simpler question would be how you determine who is shaking whose hand. Is the other person shaking your hand, are you shaking the other person's hand, or are you both actively shaking each others' hands? Just looking at both hands will not answer your question, you need to check whether or not you have initiated the movement. And this causality between initiating a movement and seeing or feeling your limbs move is the key to discover what limbs are yours. Hence, we need to look at the *frontal lobe*'s primary motor cortex, which is directly adjacent to the somatosensory area of the parietal lobe.*primary motor cortex.*

> **FRONTAL LOBE** · The *frontal lobe* of the neocortex deals with running a simplified simulation of the world. It provides us the ability to plan and evaluate actions and their future impact. The somatosensory area of the parietal lobe is directly adjacent to the frontal lobe.

As we have learned in Chapter 5.2.1, the primary motor cortex deals with activating muscles by sending signals to the spine. Like the somatosensory area, different parts of the primary motor cortex are specialized for different muscles in the body. It makes sense to have both areas close to each other given that the brain needs haptic information to adjust movements, and information about how our limbs move to adjust our sense of touch. For example, simply holding a glass of water requires you to adapt your grip accordingly if you feel it slipping. Or when drawing or writing something with a pen, you need to adapt your writing to the resistance of the pen and paper. Too strong and the pen breaks, too light and nothing shows on the page.

While we can control all muscles of our body with the primary motor cortex, we become effective only once we use our muscles in a coordinated fashion. This is the task of the *premotor cortex* of the frontal lobe (adjacent to the primary motor cortex). For example, when reaching for a glass, your shoulder, arm and hand muscles have to follow a specific motor program. But you are not thinking consciously about those individual steps. You think only about grasping the glass and the motor program is calculated by your brain in the premotor cortex. Other examples of programs created by the premotor cortex are hand-to-mouth movements, or defensive movements (for example, raising your hands to protect your head).

> **PREMOTOR CORTEX** · The *premotor cortex* is part of the frontal lobe and prepares motor programs to be executed by the adjacent *primary motor cortex*. It has strong connections to the *superior parietal lobe* which provides a model of the current state of the limbs.

In that regard, the *premotor cortex* is similar to the previously discussed superior parietal lobe: both integrate information into a greater whole. The former combines individual actions into motor programs, the latter combines individual sensory information into an idea of the status of the body as a whole. Unsurprisingly, there are strong connections between both parts of the brain. Scientists have found that the creation of the body schema begins in the womb when babies are moving their limbs.[64] Combining the feeling of the resistance (somatosensory area) when intentionally kicking (motor cortex) helps the baby to create an initial body schema based on touch. At first, we move our limbs uncontrolled, simply to test how our body reacts. Over time, we create a body schema of our arms, legs, and other body parts in order to be able to coordinate our actions.

The premotor cortex is also strongly connected to the basal ganglia. The reason for this is at any given moment, only one motor program should be executed. As the arbiter, the basal ganglia make sure that we focus our movements on one program and not switch back and forth between different motor programs. It accomplishes this by

---

[64]Whitehead, Meek, and Fabrizi, 2018.

inhibiting signals of competing motor programs. Without this focus, we could end up trying to, for example, walk forward and backward at the same time.

Also adjacent to the primary motor cortex are the *supplementary motor areas* (SMA). They are adjacent to each other and are located at the top of the frontal lobe, one in each hemisphere. This closeness relates to their function of stabilizing the body and coordinating both sides of the body. For example, rock climbing has been correlated with strong activation of the supplementary motor area.[65]

5.4.3 Alien Hand Syndrome

A lack of synchronization between the hemispheres (and therefore the primary motor cortex) can lead to a condition called "Alien Hand Syndrome" (AHS) in which the hand will sometimes reach for objects without the subject wanting to do so, even with the effect of self-harm. Patients sometimes have to physically wrestle with their "alien hand" in order to control it. For example, it can happen that the intent (stemming from the right hemisphere) of moving the left hand does not reach the left hemisphere. If, in addition, the left hemisphere observes that the left hand is moving, it concludes that an outside ("alien") force must have moved the hand. The left hemisphere does not associate the original intent of the right hemisphere with the observed movement because it never got the information that the right hemisphere wanted to move the hand. In that regard, AHS is similar to reflexes or involuntary muscle spasms due to exertion—we can observe our body moving, but there was no intent to do so. The difference between AHS and reflexes or muscle spasms is that the former requires processing in the brain while the latter involves only the spine or the muscles themselves (see Figure 5.23).

---

[65]Carius et al., 2020.

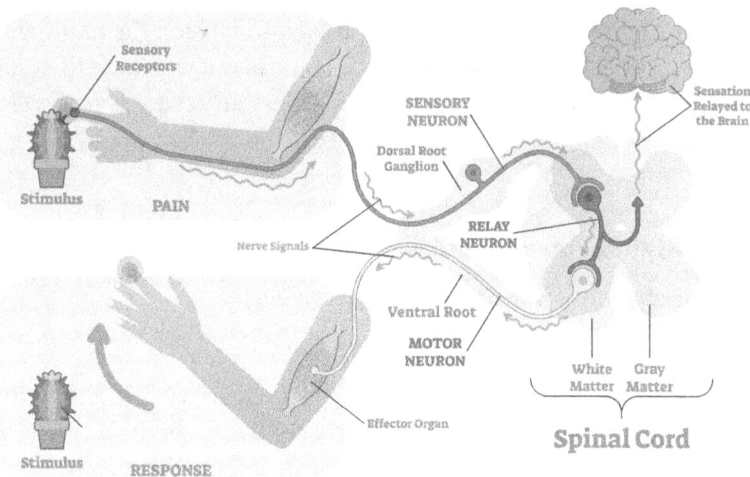

Figure 5.23: An illustration of the pain reflex involving control only in the spine (image source: Shutterstock).

It seems that to correctly identify something as self, the brain requires two independent sources of information. With what we have learned about AHS, we can conclude that these sources do not have to be the visual sense and the sense of touch, but can also be a combination of our visual sense and the internal acknowledgement that a movement was initiated by our brain and not some outside force. The following internal dialog illustrates how the hemispheres combine their information to understand what is going on.

- **Right hemisphere:** I want to reach that glass of water.

- **Left hemisphere:** Understood, go ahead.

- **Right hemisphere:** *Moves left hand.*

- **Left hemisphere:** OK, I see the hand moving to grasp the glass of water. I knew this was going to happen, all is fine.

In someone with communication problems between the two hemispheres, the left hemisphere knows as little about why an individual moved his left hand as that individual understands why *someone else* moved her hand.

- **Right hemisphere:** I want to reach that glass of water.

- **Left hemisphere:** I cannot hear you.

- **Right hemisphere:** *Moves left hand.*

- **Left hemisphere:** OK, I see the hand moving to grasp the glass of water. This cannot be caused by this brain as the right hemisphere never reported to me its intent. There must be an outside force moving the hand.

He could still try to predict his hand's behavior, but without knowing the internal state of his other hemisphere, it would be only a guess, about as accurate as predicting someone else's behavior.[66] It is similar to people suffering from dissociative identity disorder (also called multiple personality disorder). They report a feeling that their body became "possessed," that they experienced sudden impulses or strong emotions not under their control, and that they became a passive observer of their actions.[67]

AHS is actually an umbrella term for a number of similar conditions with different causes. This is because to create the body schema, many different parts of the brain are required. If there are issues in this network (for example, a damaged connection between both hemispheres), the brain misses important information to build a proper body schema. We will discuss the interaction between both hemispheres in more detail in Chapter 6.2.

---

[66]Cleeremans, 2011.
[67]Publishing, 2013.

5.4.4 Phantom Limb Pain

Now, what about the opposite of creating a body schema? What happens if our body is severely injured, and our mind's body schema no longer aligns with our body? For example, there is the condition called "phantom limb pain" where the patient suffers pain in a limb he has lost. This pain is located *outside* of the body, where the original limb was. It is not enough for the patient to simply look at the place where the limb should be and convince himself that the pain should not be there. The experience is a conflict between the learned body schema and the available visual data.

Phantom limb pain can be seen as the opposite of the alien hand syndrome. Sufferers of phantom limb pain want to *lose* the part of their body schema that resembles their formerly healthy (but now painful and missing) limb. Sufferers of AHS want to *regain* mental control of their limb. The former want to modify their body schema to reflect that the limb is lost; the latter want to modify it so that it contains the limb. Sufferers of phantom limb pain often report that their missing limb is in a clenched state and that they are unable to, for example, open their fist to get rid of the pain.

One way to treat the condition is to suppress the pain by medication. Another way would be by attaching a transplant from a donor or a prosthetic limb. A third alternative targets the body schema itself. But unlike a baby with healthy limbs in the womb, someone who is missing a limb cannot touch anything with their phantom limb. Instead, the patient's healthy limb is placed beside a mirror which is placed between the remaining and the missing limb so that from the side, it looks like the patient still has two working limbs. He is advised to repeatedly clench and unclench his healthy limb as well as his missing limb (that the brain thinks he still has and thus causes pain). After a while, by observing how the missing limb supposedly *does* work, the brain notices that the patient somehow regained the ability to move his phantom limb. Studies with patients following a six-week mirror therapy program showed that some patients experienced a significant reduction in phantom limb pain.[68]

This phenomenon can also be used to *extend* the body schema of a healthy person. Ask a volunteer to hide his hand behind a screen. Then place a fake rubber hand in front of him. Then, start stroking both the hidden hand and the fake hand with a brush. After a minute or two, the fake hand is integrated into the body schema of the volunteer. The person starts to feel that the fake hand is part of him, and when asked to close his eyes and point to his hand, he will point to the fake hand, and will even experience pain or shock if the fake hand is "injured." This is called the "rubber hand illusion."[69]

> **RUBBER HAND ILLUSION** · The *rubber hand illusion* can be evoked by brushing both a fake rubber hand and the real hand of a human participant. If only the fake rubber hand is visible, the brain will assume it is the real hand and incorporate it into its body schema.

This confirms what we have speculated earlier: creating a body schema, modifying a broken body schema, and extending an existing body schema all have in common that they rely on two sources of information. The brain of a baby in the womb is learning to connect the action of kicking with the feeling of the resistance of the kick. With the phantom limb pain, the brain is connecting the intent of clenching and unclenching the hand with visual information of the mirror image. With the fake rubber hand, the brain is connecting the feeling of brush strokes on the hand with the visual image of the fake rubber hand. We can find this idea of relying on multiple sources to establish our subjective idea of the world in other places as well. For example, think of rumours you hear and how much more believable they become when you hear the same piece of information from several of your friends.

---

[68] MacIver et al., 2008; Ramachandran and Rogers-Ramachandran, 1996.
[69] Botvinick and Cohen, 1998.

What we can do with mirrors and rubber hands, we can do even better with virtual reality and whole bodies. For example, in one study, people were put into a virtual reality (using headsets) where each participant could see a virtual representation (an avatar) of his own body from two meters away. Then, each participant was shown a hand stroking his or her avatar, while at the same time, the researchers stroked the actual body the participant. Just as with the fake hand, the participants reported that they started feeling that the avatar somehow *belonged* to their own bodies. This was demonstrated by removing their headsets, blindfolding them, moving them to another spot in the room, and then asking them to return to their original spot. In those cases where the real and virtual backs were stroked in sync, the participants overshot their original location, moving closer to the position of their avatar.[70] These kinds of out-of-body experiences can also occur without an elaborate experimental setup. Some people report that they have experienced the world from a location outside their physical body, probably caused by a miscalculation by the brain.[71] Besides the treatment of phantom limbs, other applications of virtual reality exist in therapies for amputees to improve motor control of prosthetics.[72]

Similar to how the brain learns to create a body schema to make use of the body, it also learns how the brain itself works. As it can better control the body with the body schema, it can better control itself with a "brain schema" (also called "theory of mind") which we will discuss in Chapter 5.5.

---

[70]Lenggenhager et al., 2007; Miller, 2007; Ehrsson, 2007.

[71]Arzy et al., 2006.

[72]Marasco et al., 2018.

5.5 The Theory of Mind

How does the brain determine what it thinks versus what other people think?

We can observe people, but we cannot look directly into someone's mind. All we can do is to infer another person's intentions and thoughts based on our experiences with and observations of them. For that, we are using what is called the *theory of mind* network of our brain. In this section, we will discuss the individual brain parts that make up this network and how it can support our ability to put ourselves into the shoes of others (empathy) and to recognize ourselves in the mirror.

> **THEORY OF MIND** · The concept of *theory of mind* refers to the ability to imagine what other people (or animals) are thinking and what they know. This provides a significant advantage for hunting (predicting whether the prey can see or otherwise sense you), or social interaction (teaching, trading, and even lying, etc.).

The theory of mind network integrates information that has been processed in the ventral and dorsal streams in the temporal and parietal lobes. The first steps in this network are the *left and right temporoparietal junctions (lTPJ and rTPJ)*. There, both the ventral and dorsal streams are combined to determine what is going on and where it is happening *relative to you.*[73]

> **TEMPOROPARIETAL JUNCTION** · The *temporoparietal junction* is a brain area between the temporal lobe and the parietal lobe, integrating data from the thalamus, as well as the limbic system, and the visual, auditory, and somatosensory systems. Combining the "what" and the "where" information, it creates a distinction between "self" and "other."

---

[73] Mishkin and Ungerleider, 1982.

While the idea of having two separate streams of information is helpful to understand what is going on, research has revealed that their relationship is more complex.[74] There are interactions between both streams before they reach the TPJ.[75] For example, if the temporal lobe (ventral stream) recognized a cow in the sense data, the parietal lobe (dorsal stream) could use that information to look for the cow down on the grass first instead of up high in a tree. If the parietal lobe has determined that there is something sitting high up in a tree, it is probably no cow—information the temporal lobe could use to refine its own processing.

As we have seen previously, the function of some brain parts can be explained by what the surrounding brain parts do. Specifically, the area of the TPJ is in the middle of the angular gyrus, supramarginal gyrus, and Wernicke's area (and in its mirrored homologous area in the right hemisphere). With this in mind, we can predict the functionality of the left and right TPJ.

The **left TPJ** (lTPJ) contains the **Wernicke's area** (mapping meaning to words, creating sentences), the **left angular gyrus** (written language, positioning of letters, words, and numbers), and the **left supramarginal gyrus** (phonological memory, extension of limbs, and tool use). The **right TPJ** (rTPJ) contains an area that is analogous to the lTPJ's **Wernicke's area** (dealing with subordinate meanings of ambiguous words), the **right angular gyrus** (spatial relationship of self to the external world), and the **right supramarginal gyrus** (interpreting postures and gestures of other people and inferring people's emotional states).

Taken together, the TPJs' responsibilities could be summarized as identifying oneself, differentiating "self" from "other," and inferring and interpreting one's own and the other person's emotions, inten-

---

[74]Sheth and R. Young, 2016.
[75]Schenk and McIntosh, 2010.

tions, and thoughts. While this network also supports the learning of the body schema, it is thought that the left and right TPJ, together with parts of the prefrontal cortex (and connecting structures in between) are especially involved in forming the theory of mind network.[76]

5.5.1 The Role of the Prefrontal Cortex

For the theory of mind, the prefrontal cortex also plays a significant role. Its basic functionality of counterbalancing other parts of the brain discussed in Chapter 5.2 evolved over time to allow for more complex behavior. This is reflected in a number of abilities we can observe in our own thinking:

Object permanence. Just because your senses no longer report the existence of an entity does not mean that the entity has ceased to exist. A tiger hiding behind a tree still exists, yet causes none of your senses to produce a signal. Without your prefrontal cortex, the tiger would literally vanish and you would have to rely on more primitive systems like the hormonal system (fear) to help you to remember that you are in danger. With adrenaline in your blood, you would keep running instead of taking a break once the tiger is out of sight. In this context, the prefrontal cortex can be understood as running a simulation of the world, similar to the way the superior parietal lobe runs a simulation of the internal state of the body, or the supramarginal gyrus continuously updates the body schema. This simulation allows you to mentally track objects outside of your current sense perception (*object permanence*).[77] Human children have to learn this ability, which is often incorporated in games like "peekabo" or "hide and seek."

---

[76]Schuwerk et al., 2016.

[77]Baird et al., 2002.

> **OBJECT PERMANENCE** · The ability of *object permanence* allows us to track predicted positions or movements of objects even after they have vanished from our field of view. By running a simplified simulation of the world, we are aware that a tiger that has jumped behind a tree is still there.

Reward delay and focus. To focus on one activity, the brain needs to suppress distractions. To accomplish this, the prefrontal cortex generates signals to *suppress* other signals. For example, reading a book in a public place requires you to tune out what is going on around you. Without the prefrontal cortex, which helps you to suppress acting upon something capturing your attention, you would act out whatever is related to objects in front of you. In patients with frontal lobe damage, this is called "utilization behavior."[78] Such patients have a reduced ability to resist anything that captures their attention and they have an urge to execute whatever action is connected with that particular item. This could be a pen to write something, or a smartphone demanding attention and stimulating the user to consume its virtual candy (checking on social media). The decision would be made by the rest of the brain without any regard to risks, your goals, or social rules (see below). Similarly, animals with a limited prefrontal cortex will eat whatever snack or play with whatever toy you present them. Training animals to do otherwise requires adding, for example, a social component of obedience to train a dog to resist eating any food that is available.

Memory. Another way of looking at the prefrontal cortex is as a "traditionalist committee" in the context of the neural committees. By making memories, our own past experiences can become advocates for a specific behavior. You might have the urge to pick a beautiful rose, but your memory tells you that it is connected with a negative experience (the pain from touching its thorns). So, you suppress the immediate reaction of grabbing the rose and first examine if and where you can touch its stem.

---

[78]Shallice et al., 1989.

Strategic thinking. Setting a goal in the future (planning) is a form of suppressing other actions. For example, when visiting downtown, you might want to calculate a sequence of shops you plan to visit and where to eat lunch. Being able to plan for the future also allows you to make predictions about the future by tracking the position and state (e.g., there are two coffee shops downtown but only one is open) of objects, animals, or even the mental state of other people with you (my friend will want coffee).

Long-term goals. If you see a candy bar behind a fence, instead of trying to walk directly to the candy bar (and bumping into the fence like a fly against a window), you take your initial motivation (the candy bar) but suppress the immediate idea of walking directly toward it. Instead, you analyze the situation, develop a strategy, and walk around the fence. A fly bumping against a window does not have this ability.

Risk management. If a tiger is waiting on the other side of the fence, neither memory nor a strategy can help you to get to the candy bar. Instead, the prefrontal cortex analyzes the situational risk and suppresses your desire to get to the candy bar in the first place. Once the perceived danger is gone, the prefrontal cortex sends the signal that the situation is safe. This is put forth as an explanation for obsessive compulsive disorder (OCD). Supposedly, this signal is blocked or too weak in sufferers of OCD who might then be unable to stop, for example, washing their hands.[79] Indeed, experiments show that people with OCD have less activation in their prefrontal cortex.[80] Some people show OCD-like symptoms regarding their home, taking extra time to visually confirm that the appliances are off or unplugged. This way, they are supporting the prefrontal cortex with a stronger "safe" signal using sense data coming from the parietal lobe.

---

[79] Apergis-Schoute et al., 2017.
[80] Hirosawa et al., 2013.

Processing emotions. In pre-historic times, emotions were crucial to our survival. For example, getting frustrated or angry allowed our ancestors to mobilize additional strength to push a boulder, break a branch, or wrestle an opponent. Likewise, fear and anxiety might have helped our ancestors to have more energy to run away (flight behavior). With the complexity of social interaction, our ancestors were also faced with situations that could not be solved by direct action. Their prefrontal cortex had to learn to reframe the emotional evaluation of a situation in a positive way or redirect frustrations into overcoming the problem. For example, imagine one of our ancestors is very hungry. Instead of stealing food from another member of the tribe and risking a fight, he could suppress his hunger and try to negotiate and trade.

If the underlying frustration remains unresolved, the prefrontal cortex can still find a way to cope with the emotions by, for example, humor in the face of adversity or loss; seeking emotional support from others; denial; escape (including drugs and self-medication); disengagement or dissociation; avoidance; or blame (other people or oneself). Negative maladaptive strategies tend to be more successful in coping in the short-term but not in the long-term as they are not addressing the underlying issue, and thus the actual conflict remains. For positive coping strategies, the prefrontal cortex must be able to construct a different scenario and present that to the amygdala (an imagination, see Chapter 6.5.2). Disorders like PTSD are basically the opposite of positive emotional coping. In people with PTSD, traumatic memories are reactivated and they emotionally relive those moments through their thalamus and amygdala.

Self-generated control. The prefrontal cortex also allows direct self-generated (meaning that there needs to be no external stimulus) control of your skeletal muscles. Life without the prefrontal cortex could be imagined as being in the gym, and the only way you could do any exercise would be if someone scared you in order to cause you to jump or run. Likewise, your eyes would be controlled by reflexes and would move only as a reaction to another moving object within your field of vision—sufficient to duck, but not enough to actually make decisions that are more than just *reactions* to your immediate environment or your inner state.

Dealing with uncertainty. Anterior to the premotor cortex is a prefrontal cortex area that deals with uncertainty.[81] Even with the decision mechanisms like the neural committees using an evolutionary algorithm to reach a decision, internal conflicts can persist. Sometimes, you end up with choices you cannot analyze further and this might lead you to think in circles. If I offer you $100 if you throw a die and get a 1, but you have to pay me $20 if you do not throw a 1, should you accept the deal? What is more important for you, the small chance of a large win, or the higher chance of a small loss? In this case, this part of the prefrontal cortex can resolve the conflict and at least make a decision (for example, based on probabilities).

Breaking addictions. By developing habits (strategies), you can overcome the addictive behavior promoted by competing brain parts. This is particularly challenging if the addictive behavior itself damages the prefrontal cortex, as is the case with alcoholism.[82]

Speech. *Broca's area* is part of the left side of the prefrontal cortex and is anterior to the parts of the primary motor and premotor cortex that deal with the face, tongue, and larynx control. Broca's area is connected to Wernicke's area, which is responsible for forming sentences, while Broca's area is responsible for translating those sentences into actual spoken words. Broca's area is even activated when not actually moving any muscle, which makes speaking or singing "in your head" possible.

---

[81] Volz, Schubotz, and Cramon, 2005.
[82] Nakamura-Palacios et al., 2014.

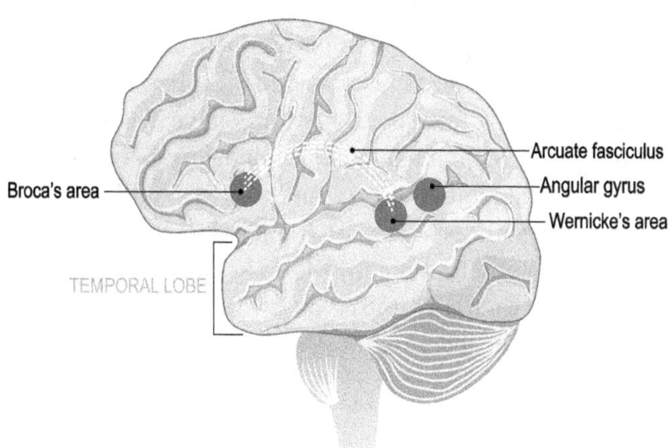

Figure 5.24: Production of language requires Broca's area in the frontal lobe to initiate and produce movements, as well as Wernicke's area to translate between actual words and meanings. If either area is damaged, this can lead to aphasia (image source: Shutterstock).

Aphasia is an impairment that can be caused by damage to the prefrontal cortex (see Figure 5.24). In such a case, the affected person still knows what he wants to say, but he is significantly hindered in his ability to find the words to express his thoughts. The analogous area on the right side of the prefrontal cortex is involved in non-verbal communication like gesticulation, rate, rhythm, and intonation of speech, as well as facial expression.[83]

> **BROCA'S AREA** · *Broca's area* is a brain part located in the left side of the frontal lobe and connected to Wernicke's area. It is responsible for the *production* of speech. Damage to Broca's area leads to a person unable to find the words to express what he wants to say. The homologous area in the right hemisphere deals with non-verbal communication.

---

[83]Johns, 2014.

Not all animals with a prefrontal cortex have all the abilities listed above. For example, besides primates, so far it has only been shown that dogs, cats, and a few species of birds have the ability to track objects once they are no longer visible (object permanence). Surprisingly, many predator animals do not have this ability. When prey suddenly vanishes underwater or behind a tree, they are visibly confused and unsure where it went. Their basic emotional system still works and they are still in hunting mode. But without an idea of the location of their prey, they might give up their hunt.

5.5.2 Mirror Test

Beyond tracking objects in the environment, the prefrontal cortex also tries to predict what other people know. This theory of mind ability can be demonstrated in an animal with the so-called *mirror test*. For example, we could work with a chimpanzee; let us call her Tinkerbell. A blot of coloring is secretly added to Tinkerbell's head where she cannot see it without the help of a mirror. We make sure she did not feel us putting a blot of color on her. We then put her in front of a mirror. We know that when faced with a mirror, chimpanzees first make threatening gestures at their own images but after a while learn to use their reflections for self-directed behaviors like making faces at their reflections or grooming previously unobserved parts of their body.[84] If she starts examining the blot of color, she passes the test, because the only way she could have known that there was a blot of coloring on her head was by recognizing herself in the mirror. Passing the test indicates that she is able to map her internal decisions to her actions, as well as her own movements to the mirror image's movements.

---

[84]Gallup, 1970.

To pass the mirror test, Tinkerbell needs several sources of information:

- The visual image of the chimpanzee in the mirror (using the visual cortex): "I am seeing a chimpanzee in the mirror."

- An idea of what chimpanzees usually look like (for example, other members of her tribe) to recognize that something is odd about the chimpanzee in the mirror (the blot of color).

- The information about what *initiated* the movement (with the help of the premotor cortex): "I know I am moving my arm."

- The recognition that her own movements match the movements of the chimpanzee in the mirror (with the help of the supramarginal gyrus): "I am seeing a chimpanzee in the mirror who *moves* like me."

- Further analysis of the movements with the conclusion that the cause of the mirror's image actions must be *herself*. Only she has the information of what she will be doing next. If the ape in the mirror is another ape and not herself, the other ape does not have the same information that Tinkerbell has. This requires the TPJ (labeling whose information belongs to whom) and prefrontal cortex (tracking who knows what): "I am causing the mirror image to move its arm," "Only I knew that I would move my arm," and subsequently "That must be me in the mirror."

- Finally, she notices that her internal body schema does not match the mirror image ("I have something on my head") and she starts examining the blot of color on her head.

While the mirror test seems to be a special case as mirrors are rare in the natural world, this ability to differentiate self and other helps whenever we are touching someone. For example, imagine holding someone's hand and then seeing it moving. Who or what caused the hand to move? With additional input from the motor and the somatosensory cortex, the rTPJ figures out how to integrate what you see (the moving hand) with the information from the self (did *I* move the hand or did the *other person* move the hand?) to make sense of what is actually happening. If you had to rely only on a single source of information (for example, your eyes), you would not know who is holding whose hands.

> **MIRROR TEST** · The *mirror test* evaluates the ability of an animal to recognize itself in a mirror after a researcher has secretly added a blot of coloring to the animal's body and put the animal in front of a mirror. If the animal starts investigating the blot of color on its own body instead of on the mirror image, it passes the test (because that indicates the animal realizes it is the same creature that it sees in the mirror).

Besides primates, also dolphins, Eurasian magpies, African elephants, and possibly giant manta rays pass the mirror test.[85] Dogs and cats seem to be at the very edge between animals that can recognize themselves and those that cannot. It is important to note, though, that the mirror test relies heavily on an animal's vision, while the main sensory focus of dogs and cats is smell. Maybe someone will design a "smell mirror" test in the future that a cat or a dog might be able to pass. While dogs can recognize their own smell, it is unclear how to test whether they do anything with that information.[86] For example, we do not know if they change their behavior (like their diet) when they smell signs of illness—and even that could be interpreted as an instinct rather than actually recognizing themselves.

Another way to demonstrate the ability for a theory of mind is the "Sally and Ann task" experiment.[87] Here, "Sally puts a marble into a basket. While Sally is away, Ann quickly puts the marble into another box. Then, the subjects are asked at which location Sally

---

[85] Ari and D'Agostino, 2016.
[86] Horowitz, 2017.
[87] Wimmer and Perner, 1983.

will search for the marble. Hence, the participants have to perform a shift in mental states and breach with their former expectation (true location) to understand Sally's false belief of the location of the marble. This involves inhibiting their intuitive response of naming the true location."[88] In other words, if they answer that Sally will look into the box where Ann put it, they have failed to demonstrate having a theory of mind. Sally could not have known that Ann put the marble into the box.

Yet another situation in which we need the theory of mind is when we want to explain something to someone. For this, we need to have the ability to know what someone else knows. This is something children have to learn while their prefrontal cortex only fully develops throughout adolescence until around the age of 25 years.[89] For example, when asked what she is doing, a child might tell her father, "I'm playing with this" and then hold up a toy car, even if this conversation occurs over the phone. This is because the child, at her age, cannot consider that the father does not see the toy car. To solve the misunderstanding, the child needs to understand the other person's perspective and add additional information. The child would need to use her prefrontal cortex to figure out what the father knows about her situation, and her TPJ to differentiate her own knowledge about the situation from her father's.

---

[88] Krall et al., 2014.
[89] Arain et al., 2013.

5.5.3 Empathy

The rTPJ enables us to see the world from another person's point of view. For the calculation of the rTPJ, the point of view (our own or that of other people) is just another input variable. This architecture allows us (with input from the prefrontal cortex) to have sympathy and empathy for other people. Just because we see a person being angry or sad does not make the person unsympathetic or sympathetic—it depends on the *context* that the rTPJ and prefrontal cortex provide (background information, body posture, facial expression, pitch, tone, etc.).

A theory for *lack* of empathy and sympathy, especially in criminals, is that the processing ability of their rTPJ is underdeveloped or malfunctioning. Experiments have shown that there is significant activity in the rTPJ when having to decide whether something is moral. In order to make that call, you have to include other people and situations into your judgement.

The connection between the rTPJ and morality was confirmed in a study where the participants' activity of their rTPJ was disrupted with magnets and then the participants were asked to read several accounts of a situation and judge whether the protagonist ("agent") acted morally.[90] The scientists' conclusion from the study was that we form moral judgments based on a combination of the agent's intentions and the actual outcome. If our ability to judge an agent's intentions is disrupted, we put more emphasis on the outcome. To illustrate the differences between intentions and outcome, let us examine one of the stories that was presented to the participants, which was about Grace ("agent") and a white powder she put into her friend's coffee (see Figure 5.25):

---

[90]L. Young et al., 2010.

	Neutral outcome	**Negative outcome**
Neutral belief	Grace thinks the powder is sugar. It is sugar. Her friend is fine.	Grace thinks the powder is sugar. It is toxic. Her friend dies.
Negative belief	Grace thinks the powder is toxic. It is sugar. Her friend is fine.	Grace thinks the powder is toxic. It is toxic. Her friend dies.

Figure 5.25: Four examples with different outcomes and initial beliefs about the situation. Depending whether or not the rTPJ of the participant in the study was disrupted, different moral judgements about Grace were made.

- **Unsuccessful attempts to do harm** (outcome neutral, belief negative). In the first variation of the story, Grace puts white powder from a container marked "toxic" into the coffee of her friend. The container was mislabeled (it was actually sugar) but Grace had to have assumed that she was putting toxic powder into her friend's coffee. In the experiment, disrupting participants' rTPJs while they were thinking about judgment of such intended but failed acts to do harm led to lenient verdicts. They judged Grace depending on the outcome (whether or not her friend died), not her intentions (whether she knew it was toxic powder).

- **Well-intended but harmful acts** (outcome negative, belief neutral). In the second variation of the story, someone replaced the contents of a sugar container with toxic powder without Grace's knowledge. When Grace put the toxic powder into her friend's coffee, she did not know she was poisoning her friend. Again, the particpants' rTPJs were disrupted while they were thinking about a judgment, and ended up with severe judgements. They judged Grace depending more on the outcome than her intentions.

As discussed in Chapter 5.4, possible therapies for real-life violent offenders might include a virtual reality simulation where they see themselves from the eyes of their victims. For example, researchers of the University of Barcelona have developed a virtual reality system so that men who committed a domestic violence crime can experience the act from the victim's point of view. The study[91] compared a group of violent offenders with a control group. The former group showed a significantly lower ability to recognize fear in a woman's face. After their virtual experience, participants in the study showed improved scores in a test where they had to guess, by looking at a picture of a person, the emotions that person is experiencing. Before they experienced themselves from their victim's perspectives in the virtual reality, violent offenders "showed a bias towards classifying fearful [...] faces as happy." Using treatment with the virtual reality experience, this bias was reduced. Further questioning also revealed that violent offenders had a significant bias to responding in a manner that would be viewed favorably by others. More research is needed for an actual therapy; it is still unclear whether it is their distorted view of themselves and others that leads to a lower threshold of using violence, or if that view is a result of trying to rationalize past violent behavior.

Such a therapy could train the violent offenders' rTPJ, putting more emphasis on the thoughts, feelings, and intentions of a person when making a moral judgment. Instead of simply blaming another person for a negative outcome (and then possibly using that as a justification for violence), they could better think about whether or not the other person really had an influence on the situation. For example, someone with a weakened rTPJ might think about his partner: "I had a bad day at work. Whatever you did, the outcome was that I had a bad day. So, you are to blame." Someone with a trained rTPJ might come to a different conclusion: "I had a bad day at work. You supported me, other factors (maybe even myself) are to blame for my bad day."

---

[91]Seinfeld et al., 2018.

> **Did you know?**
>
> The approach of splitting blame into observations, feelings, needs, and requests is the central idea of Marshall Rosenberg's method of non-violent communication. It provides on way of training your emotional intelligence and creating a psychologically safe space.
>
> $\longrightarrow$ Read more in *Philosophy for Heroes: Persona*

This idea is supported by findings that the rTPJ's volume is strongly associated with an individual's altruism.[92] Stimulation of the rTPJ seems to enhance a person's ability to take someone else's perspective.[93] Better connectivity of the rTPJ also seems to result in less bias toward people outside one's social group.[94]

Beyond empathizing with other people, the ability to see things from another person's perspective is also relevant to empathizing with *one's future self*. Stimulating the rTPJ seems to have an effect on one's ability for self-control.[95] In experiments, participants were asked whether they would prefer to have a pile of money now, or a significantly larger pile of money later. With the rTPJ interrupted, they preferred the short-term win over the long-term gains. This was supported by other studies comparing empathy and impulsivity,[96] and examining the relationship between substance abuse and empathy.[97] It seems that for the brain, it does not matter if the person with whom we are empathizing is another person or our future self.

---

[92] Morishima et al., 2012.

[93] Santiesteban et al., 2012.

[94] Baumgartner et al., 2015.

[95] Soutschek et al., 2016.

[96] Pajevic et al., 2018.

[97] McCown, 1989.

> **Did you know?**
>
> There are many correlations between the activity of specific
> brain parts and an individual's personality and personal pref-
> erences such as impulsivity, empathy, directness, openness,
> etc.
>
> $\longrightarrow$ Read more in *Philosophy for Heroes: Persona*

Similarly, together with the prefrontal cortex, the lTPJ also helps an
individual to know her own (and those of others) thoughts, beliefs,
intentions, and desires.[98] For example, detecting a lie requires the
ability to recognize the conflict between another person's inner state
(what they know) and what they are expressing. Damage to the lTPJ
can impair this ability.[99] Such damage also impairs the ability to tell
convincing lies if doing so requires knowing what the other person
thinks.

Knowing someone else's internal state enhances our ability to teach
and lead other people. If the lTPJ is damaged, our brain might no
longer recognize that other people's minds can differ from our own.
In this case, we would lack the motivation to tell other people what
we have learned: we would assume that they already know what we
know. With the help of a healthy lTPJ and by using context information from the prefrontal cortex, we can track who might know
what.

To build a theory of the mind, the prefrontal cortex predicts how
people (and objects) in the environment will react and what other
people think or believe, while the TPJ connects this information to
actual entities in the real world. For example, the lTPJ is involved
in the association and memorization of names of individuals and

---

[98] Gallagher et al., 2000.
[99] Samson et al., 2004.

objects.[100] If you know that Peter likes Sally, this is just abstract information in your prefrontal cortex. It becomes concrete when the lTPJ provides the information that *you* are Peter and the person in front of you is Sally.

We are so used to trying to predict what other people are thinking that we even attribute an inner state to non-living things. Think of how people curse at non-functioning computers or cars, despite knowing that this behavior would make sense only if the computers or cars had a human-like mind. A car does not—indeed cannot—care what we say to it. Is consciousness merely the attributing of an internal state to things and people, or does it have a deeper function? This will be the topic of Chapter 6.

---

[100] Gorno-Tempini et al., 1998.

Chapter 6

Consciousness

Problems that remain persistently insoluble should always be suspected as questions asked in the wrong way.

—Alan Watts

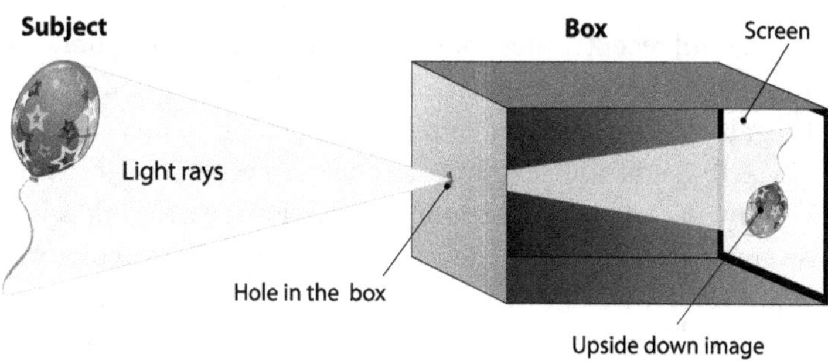

Figure 6.1: The camera obscura (or pinhole camera) is the basis for modern cameras and gives us a basic idea of how our eyes work (image source: Shutterstock).

Our intuitive understanding of how we *perceive* the world has changed significantly over time. For example, let us consider sight. In Ancient Greece, people believed that rays *emitted* by our eyes somehow allowed us to see. It took until the 10th century for someone (the astronomer and physicist Ibn al-Haitham) to demonstrate how the the pupils and retinas of our eyes actually work. His pinhole camera (see Figure 6.1) projected an inverted image through a hole (the pupil) on a screen (the retina). In the ensuing centuries, particularly with the scientific method, we have found better ways than relying on our intuition to discover how the world works.

INTUITION · Your initial evaluation of a situation is called *intuition.* It is the first thing that comes to your mind without going through conscious deliberation or reasoning.

Yet, despite advances in philosophy and with more than a century of neurobiological research under our belt, no one has been able to locate (in the brain) what we call "consciousness." Given the (superficial) similarities between human and computer intelligence (both process information and make decisions) and the amount of accumulated research in computer science, one might believe that at least a basic concept of the conscious experience has been discovered by scientists. While there have been many descriptions of human consciousness, until recently there has been no definitive philosophic or scientific answer to the question of what *precisely* consciousness is. In that regard, the challenge of defining or explaining consciousness seems to be that it is not something anyone can observe objectively. I am the only one able to report on my consciousness. While some-one else could scan my brain and say that I am seeing the color red, that scan does not explain why there is an "I" that is experiencing the sensation of "seeing red." Even more troublesome, the only way I can report on my experience is using the very thing I want to report on (my consciousness).

Now, how should we approach this problem? The best way is to just work with what we have, see how far we can get with it, then take our learnings and re-evaluate our original premises. It is the same problem we were facing in *Philosophy for Heroes: Knowledge*: ontol-ogy ("What is?") and epistemology ("How do we know?") are inter-twined. Learning more about the world helps us to refine our knowl-edge about *how we learn about the world*, which in turn helps us to learn about the world more accurately, and so on (see Figure 6.2).

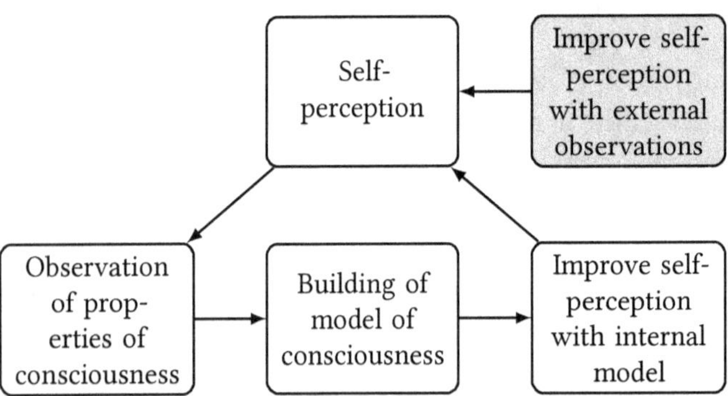

Figure 6.2: We can refine our understanding of consciousness by repeatedly re-examining our self-perception using what we have learned previously.

In Chapter 5, we have begun this iterative process by looking at optical illusions, cognitive defects, our evolutionary history, a comparison with our distant cousins, the apes, and the architecture of our brain. The more we investigate the brain, the more it seems that the brain could function on its own, without the need for a conscious experience. This raises numerous questions:

- Why do we have a conscious experience at all?

- Why does this conscious experience feel as if it is separate from the world?

- If our conscious experience is separate from the physical world, how could it influence our physical actions, and how can interactions with the physical world create conscious experiences?

- When we have a subjective experience, who or what is experiencing it?

In this chapter, we will explore these questions from several angles.

Theories of Consciousness: To build a solid definition of consciousness, we first need to consider possible interpretations of consciousness from philosophers and scientists in the past, as well as to explain consciousness in the context of evolution. Until we have defined consciousness, we will use the word as an umbrella term for anything relating to the equally vague terms of our subjective experience, will, decision-making, awareness, and the "self."

The Location of Consciousness: We know where, for example, our toes are located, but our subjective experience of consciousness only gives us a vague idea that consciousness is somewhere within us. We examine whether it arises from the brain as a whole, whether there is an observer within our brain, and whether there is a difference between the left and right hemispheres. Looking at a number of cognitive defects, we refine our idea of what it means to be conscious.

The Search for Consciousness: Revisiting the question about how consciousness could be the result of an infinite series of observers, we come up with a basic theory correlating with brain architecture. Using a loop of consciousness and central places to store memory, we can better explain the role of the prefrontal cortex, how we can imagine things, and how something like a "mind's eye" or "inner voice" is created.

Model-Building: Besides processing sense perception, the second major function of our brain is to think abstractly and predict the future. For this, we come back to the concept-building from *Philosophy for Heroes: Knowledge*, and discover how the brain learns to build models of other people and the environment.

The Attention and Awareness Schema Theory: Beyond attention and awareness, the brain also builds *models* of attention and awareness, which enable us to think about our past, present, and future selves. We conclude the book by putting together all that we have discussed to arrive at a definition of consciousness.

Building a Conscious Mind: At the end of the book, we will summarize what we have learned about the brain and consciousness into an evolutionary timeline. Our goal will be to explain how human consciousness could have developed gradually.

6.1 Theories of Consciousness

It is easy to believe that the mind and the brain are two separate things. After all, we *perceive* consciousness to be something immaterial (it is difficult to locate and impossible to touch) and non-causal (intuitively, we see ourselves as independent actors with free will). In fact, our subjective experience of the world seems far too rich and intense to ever be explained by a physical process in the brain. How can we be sure that this intuitive distinction is based in reality and does not simply stem from having separate brain parts to think about ideas (theory of mind) and things (object permanence)?

The challenge with the approach of seeing *mind* and *brain* as separate is that such an approach works fine in daily life only as long as we have a healthy brain. For example, in Chapter 5.5 we learned that Wernicke's area is the part of the brain responsible for forming words and sentences. Damage to this brain area can also affect a person's ability to find the right words or put them in correct order. Either the theory of having a separate mind and brain is wrong, or coming up with words and sentences is a function of the brain and not the mind. This could be repeated with any other brain impairment with fewer and fewer functions attributed to the mind. This apparent contradiction between the intuitive understanding and observable facts has been part of philosophical discourse for centuries.

6.1.1 History of Theories of Consciousness

Why did philosophers and scientists of the past struggle with the question of consciousness? What theories did they rely on to explain the material brain and the immaterial mind?

Plato (429–347 BC) studied and wrote about the perceived differences between the material world (our body, including our brain) and the immaterial world (our subjective experience of the world). He called his approach the "theory of forms." Plato's position was that concepts or ideas were the basic building blocks of reality and the material world was merely a shadow or imitation of this ideal world. As such, Plato saw consciousness as part of the ideal world. He thought that a person's consciousness creates mental images of concepts and ideas that are only imperfect once an individual tries to manifest those mental images in the real world. For example, there is the concept of an apple, and there are actual (imperfect) apples. We can imagine the ideal concept of an apple, but when actually trying to draw it or grow it, our apple is merely an imperfect copy of the ideal apple. Likewise, there are ideal forms of "us" and there are our actual bodies—imperfect representations of "us." This view that mind and brain are separate is called "mind-brain dualism" or also "mind-body dualism" (with "body" including our brain).

> **MIND-BRAIN DUALISM** · The philosophical view of *mind-brain dualism* (or also *mind-body dualism* with "body" including our brain) states that what we call the (immaterial) mind is separate from the material world (the body, including the brain).

Plato used the *allegory of the cave* to stress his point that we could exist in an allegorical cave, just watching shadows of the real world. Plato thought that the ideal form of humans existed both before our birth and after our death. This ideal form is connected with the real world only temporarily during our life on Earth. Plato's disciple Aristotle (384–322 BC) disagreed with Plato's approach of the *pri-*

macy of consciousness (see *Philosophy for Heroes: Knowledge*) where ideas about an entity (Plato's Forms) are more important than the entity itself. While Aristotle also believed in a distinction between the material brain (part of the body) and the immaterial mind, he did not place one above the other. In his view, the mind was a property of the brain and died with the brain.

For Aristotle, an apple had properties that made it an apple, while for Plato, there was a separate realm of an ideal apple of which the physical apple is just a representation. Still, Aristotle believed that conceptual thought was separate from the physical brain, as he could not see how something material as the brain could process concepts which in themselves are essentially limitless. In his view, there must be something special about the mind as we can think about any kind of material object.

Mind-brain dualism was not unique to Plato and Aristotle, although they were among the first who examined it systematically. Abrahamic religions like Judaism, Christianity, or Islam share a view regarding the relationship between mind and brain that lies somewhere between Aristotle and Plato. *Judaism* (with its origins around 600 BC) shares Plato's basic ideas about an immaterial afterlife and a dualistic world, while *Christianity* (Jesus lived around 6 BC to 33 AD) and *Islam* (the Quran was written from 609 to 632 AD) share the belief in a bodily resurrection.

In Thomas Aquinas' (1225–1274, an influential Catholic theologian) discussion of mind and brain (or soul and body), he does not see them as two separate entities interacting with each other, but instead as one. At the same time, he still argues for a separate, incorruptible soul. Thus, his philosophical position is somewhere between mind-brain dualism and the view that all we have is a material existence.

Later, Descartes (1596–1650) focused on the question of how the immaterial mind (or Plato's ideal Form, or Aquinas' soul) interacts with the brain. He proposed that there is a connection between the material and immaterial brain through a specific brain part. Given that it is located deep in the center of the brain, the primary candidate was the pineal gland, as it had been attributed with all sorts of mystical properties throughout history. In this theory, information would flow from the brain to the pineal gland, which communicates the information to the immaterial world where consciousness and free will supposedly sit. Having thought about the information, we (the immaterial mind) communicate back to the brain through the pineal gland, which results in an action.

To figure out where exactly to draw the line between the material brain and the immaterial mind, Descartes came up with the following thought experiment:

> I shall think that the sky, the air, the earth, colors, shapes, sounds, and all external things are merely the delusions of dreams which [the evil demon] has devised to ensnare my judgment. I shall consider myself as not having hands or eyes, or flesh, or blood or senses, but as falsely believing that I have all these things.

—René Descartes, Meditations, 1641

EVIL DEMON · Descartes' *evil demon* is a thought experiment to differentiate the immaterial mind from the brain. His "evil demon" is an entity that could make any changes to the material world without anyone being aware of such changes. Descartes' assumption was that by examining the remaining things we could rely on, we would discover what the immaterial mind is.

The demon could feed you wrong information, but—at least in Descartes' view—could not affect your ability to *think about* the information. So, maybe the colors, shapes, and sounds you are perceiving are not real, but at least the fact that you are making those perceptions is real. The reason for this rather odd approach is that Descartes wanted to examine such a question without bias. What would be left for the immaterial mind if nothing we see, hear, taste, feel, or touch could be counted on as being "real"?

Descartes argued that humans had a *non-physical mind* with consciousness (everything the demon could *not* affect directly, for example the subjective experience, the inner voice, the ability to perceive things, or the imagination), and a *physical brain* and sense organs (which could be influenced by the demon, directly or indirectly by the environment) that regulates bodily functions and provides intelligence ("What is 2+2?") and perception ("What color is the box?").

> ### Biography—**René Descartes**
>
> René Descartes was a French polymath (1596–1650) and is widely regarded as one of the founders of modern philosophy. He attempted to write about philosophy "as if no one had written on these matters before" (his words) and is best known for his statement "I think, therefore I am"—an attempt to explain the foundations of ontology ("what is?") and epistemology ("how do we know?"). Besides his philosophical writings, he had significant influence in the field of analytical geometry (the Cartesian coordinate system is named after him).

6.1.2 The Origin of Consciousness

Which of our ancestors was the first to experience consciousness?

While Descartes' distinction between the mind and brain at least provides usable definitions, it misses the explanation of *how* both of these entities would interact. But even more problematic is that it also misses the explanation of how the physical brain evolved to be able to access the non-physical mind. Mind-brain dualist philosophies generally skip this question, arguing that they must have had no consciousness at one point, and then suddenly had it at another. Religions like Christianity do not directly address this issue. It is believed that humans are different from all other creatures on Earth, having a soul which was either given to humans or with which they were originally created.

Since Descartes and Darwin, scientists were able to materialistically explain more and more functions of what Descartes called the non-physical mind. This diminished the need for an immaterial mind the physical brain could somehow access. Hence, more radical ideas that question existence itself became more popular. They proclaim that, after all, consciousness is primary, meaning that *existence is created by consciousness* and that your thoughts become physical reality. Therefore, the universe and all the matter it contained would be just the product of the underlying *actual* consciousness. Having a primacy of consciousness would address how the brain could have evolved: if consciousness is primary, then it would be the driver of evolution and, in a sense, it would mean that consciousness would have been always there, and it somehow built a brain and body around itself. It would also address how mind and brain could communicate (both mind and brain originate from the same consciousness).

While the primacy of consciousness seems to be the answer we are looking for, it comes with a lot of philosophical baggage. *How* exactly would consciousness direct evolution to create our brain? As we cannot easily resolve this issue, let us consider a more general idea. Let us assume that both the mind and brain originated from a single source and see if this helps explaining both the communication between mind and brain and the evolution of consciousness.

6.1.3 Monism

While we have established that in dualism, there needs to be some kind of communication that goes back and forth between mind and brain, this condition does not necessarily apply if mind and brain are the same, if they come from the same source and run in parallel, or if there is only our mind and no brain or body. This view is a variant of what is called *monism* (see Figure 6.3). Monist views of consciousness have existed for many centuries. They assume that reality ultimately consists of only one substance—an approach that makes the search for a connection between the material and immaterial world superfluous.

> **Monism** · The philosophical view of *Monism* is that everything that exists (including what we call mind or consciousness) can be traced back to a single fabric of the universe. The consequence of this view is that the mind must be a result of a (mechanistic materialist) process and that mind and matter are not two separate things.

While there are many different variants of monism, the main categories are:

> **Materialism** · In the *materialist* worldview, there is no separate "mind." Instead, everything can be explained by a single substance (matter).

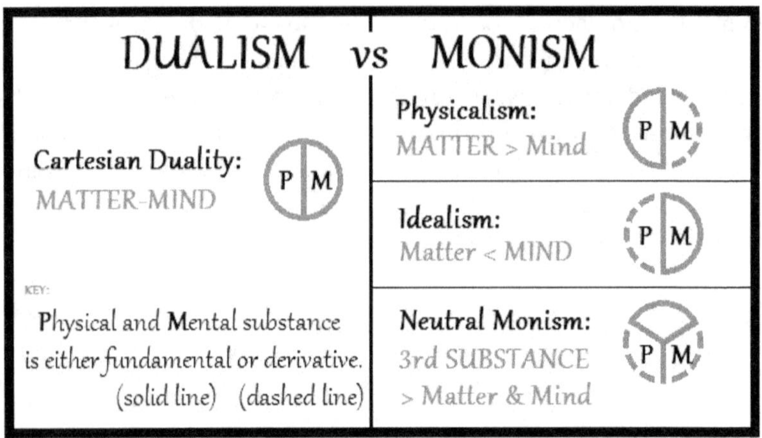

Figure 6.3: Comparison of different types of monism.

> **NEUTRAL MONISM** · In the *neutral monist* worldview, while both mind and matter exist, they both stem from a third, undefined substance.

According to Baruch de Spinoza (1632–1677), both mind and body (including the brain) are merely different "modi" of God, resulting in the same manifestation. When you move your arm, it is not your will that causes your arm to move. Instead, something else causes both the movement of the arm, as well as the experience that you are moving your arm. This is also called *psychophysical parallelism.* In this view, the mind and brain are separate, and they do not interact with each other. Instead, everything is parallel, so that it merely looks like they are interacting. This parallelism is either established by repeated Godly intervention, for example, when our body is hurt, God makes the mind feel pain (Nicolas Malebranche, 1638–1715), or by perfect starting conditions like two clocks showing the same time although they are not connected (Gottfried Leibniz, 1646–1716).

This parallelism is similar to what we discussed in *Philosophy for Heroes: Continuum* about the "spooky action at a distance" when it comes to quantum mechanics. Let us say I split a coin horizontally into two and give you one half without either of us looking at it and then I leave on a journey to the stars. While my part of the coin is then light-years away, it is still connected with your part of the coin (if I look at the coin, I immediately know which side of the coin you have) because both coin parts share an initial action.

Parallelism is basically materialism but with the added explanation of where our subjective experience comes from. If we can explain our subjective experience with materialism, we can safely ignore parallelism and just focus on materialism.

> **SUBJECTIVE IDEALISM** · In the *subjective idealist* worldview, there is no such thing as "matter." Instead, everything is but perception, mind, or "consciousness," and nothing exists but human minds and gods.

Gottfried Leibniz (1646–1716) described a variant of subjective idealism (*pluralistic idealism*) which states that the world is the product of many individual minds that generate a virtual world, similar to an online multiplayer computer game.

> **INDIAN IDEALISM (VEDANTA)** · In the *Indian idealist* worldview, there is but a single consciousness and our experience of the world as separate beings or consciousnesses is an illusion.

Technically, *Buddhism* and *Objectivism* are monist world views, too. We can use them as sources of inspiration as they already include something that resembles the concept of attention. Ignoring their special positions concerning focus and free will (we will pick up the topic of free will in the next book, *Philosophy for Heroes: Persona*), Objectivism most closely resembles materialism, while Buddhism most closely resembles idealism, so we will focus on those instead.

OBJECTIVISM · *Objectivism* (founded by Ayn Rand, 1902–1982) is a monist philosophy that recognizes a distinction between the brain and a "prime mover" that has a power of whether to focus the brain or not.

BUDDHISM · *Buddhists* (Buddha lived 563–483 BC or 480–400 BC depending on the source) believe in a difference between brain and consciousness, with consciousness being compared to a light that shines on thoughts. The mind dies with the bodily death while "you" are reborn into a new being with no memories of your previous life.

With this discussion in mind, we can easily see how religions (and of course philosophies) are based on different interpretations of those basic philosophic questions—Monism versus Dualism, rebirth, and primacy of consciousness (see Figure 6.4):

Name	Primacy of consciousness	Dualism	Afterlife
Platonism	yes	yes	yes
Judaism	yes	yes	yes
Islam	yes	yes	yes
Christianity	yes	yes	yes
Aristotelianism	no	yes	no
Objectivism	neither	neither	no
Idealism	yes	no	differs
Buddhism	neither	neither	rebirth
Parallelism	no	neither	differs
Materialism	no	no	no

Figure 6.4: Comparison of different religions and philosophies based on their position on questions of primacy of consciousness, mind-brain dualism, and afterlife.

6.1.4 Quantum Mind Hypothesis

Does human consciousness exist in the quantum world?

Instead of trying to see brain and mind, materialism and idealism, or ontology and epistemology as irreconcilable, it helps to see them as two sides of the same coin. An analogy would be the hardware and software of a computer. Is a computer a series of connected modules, or is it program code that gets executed? Arguing that computers are just program code sounds as silly as arguing that we are just consciousness. The same goes with arguing that computers are just hardware. Instead, it helps to see them as an integrated system: individual modules through which data flows. Hence, our approach will be to identify the role of the individual brain parts and connect them together to explain how information flows through them.

In recent years, a group of hypotheses that gained popularity proclaimed that consciousness cannot be explained with classical mechanics. Their authors are proposing the existence of a "quantum mind" which could bridge the gap between mind and brain through quantum mechanics. The quantum mind would reside in a separate quantum world and would have to determine the state of the brain at a quantum level (the quantum brain) to access the sense data from the sense organs. Moving in the other direction (quantum mind to brain to muscle), the quantum mind would have to affect the macroscopic (classical) brain, which in turn could cause the body to act based on the command given by the quantum mind (see Figure 6.5).

> **Quantum mind** · The term *quantum mind* refers to a collection of theories that consider quantum mechanics as the basis of consciousness.

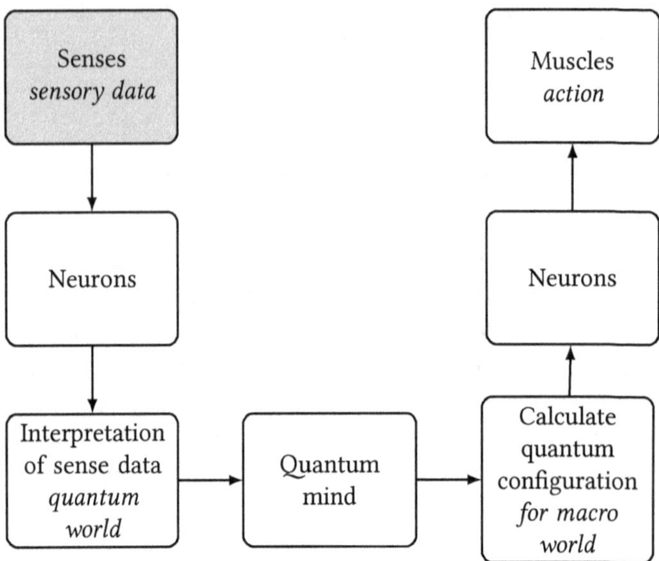

Figure 6.5: The proposed architecture of a quantum mind.

The idea of a quantum mind seems plausible as both consciousness as well as quantum mechanics are seen as equally difficult to understand. But the question remains that if consciousness is based on a quantum mind, how would it work? After all, the reason to move the mind into the physical world is to be able to explain how it works.

Looking at the actual research, we quickly discover that the answer depends on the specific *interpretation* of quantum mechanics that is being used. In the scientific community, some twenty different interpretations of "quantum mechanics" are recognized, all being mathematically correct according to quantum theory. They involve different philosophical assumptions about causality and the role of the observer (see *Philosophy for Heroes: Continuum*). If you read something along the lines of "scientists prove that the world is created by thoughts," that means only that the scientists have used a proven mathematical theory and built their (unproven) interpretation on top of it. Just because the interpretation is non-contradictory does not mean it is true.

One theory assumes that the universe splits into many different versions of itself with each decision you make. But this *many-minds interpretation* is not proven; it is merely an *interpretation* of what the results of the underlying (and mathematically sound) quantum theory *could* mean.

> **Many-minds interpretation** · The *many-minds interpretation* of quantum mechanics is similar to the many-worlds interpretation, in which the universe splits into infinite universes. In the many-minds interpretation, the split of the universe happens with each thought for each individual brain, instead of with each measurement (as in the many-worlds interpretation). Consequently, one's consciousness splits into many consciousnesses whenever a decision was made.

An ontologically less wasteful theory was the original interpretation of quantum mechanics, the *Copenhagen interpretation* (also see *Philosophy for Heroes: Continuum* for a detailed discussion).

> **Copenhagen interpretation** · In the *Copenhagen interpretation* of quantum mechanics, an observer is required for the wave function to collapse. Without observation, the wave function never collapses and never becomes a particle. While the interpretation does not mention consciousness as such (measurements by a device are observations, too), it raises the question of who observes the observer, resulting in an infinite loop.

The *von Neumann-Wigner interpretation* of quantum mechanics also assumes that entities exist only when they are observed. But it breaks the infinite loop by arguing that the quantum mind is the final observer, and that the final observer (the quantum mind) does *not* need to be observed itself in order to exist (see Figure 6.6).

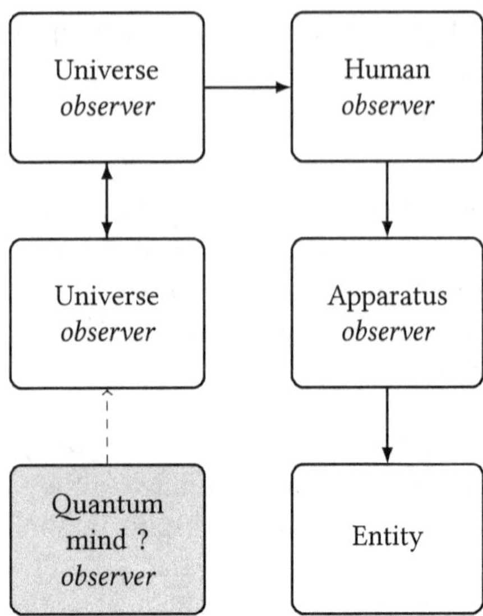

Figure 6.6: The von Neumann-Wigner interpretation addresses the Copenhagen interpretation's observer problem by arguing for a final "quantum mind" observer.

VON NEUMANN-WIGNER INTERPRETATION · The *von Neumann-Wigner interpretation* of quantum mechanics tries to solve the loop of the Copenhagen interpretation by stating that consciousness is outside of the quantum world and is the final observer that collapses the wave function. For consciousness itself to exist, no observation of consciousness would be needed.

While these interpretations of quantum mechanics involve consciousness to some degree, none of them gives us any deeper insight into what consciousness could be or how we can describe it. Biology and physics seem to be pointing at each other to solve problems relating to consciousness. Hence, we focus on the connection between the quantum world (physics) and the macro world (biology). If we can determine how both worlds communicate with each other, maybe we can answer the question about the nature of the quantum mind.

6.1.5 A Janitor's Dream

How could something existing in the quantum world have an effect on the macro world?

In a way, the supposed influence of quantum fluctuations of a quantum mind on our thinking could be compared to the impact of genetic mutations on evolution. In *Philosophy for Heroes: Continuum*, we made the point that mutations are not the drivers of evolution, and they merely provide *variety* in the gene pool. The actual driver of evolution is the *selection* of the best adapted individuals for procreation of the next generation. For example, eyes did not develop through a series of mutations. Instead, minuscule differences in the eyes (or precursors of eyes) provided a small evolutionary advantage (like providing images with a higher resolution). Similarly, in the brain, signals pass through several layers of filtering (selection!) while noise (a mutation) is getting filtered out. This happens by focusing on patterns in the incoming sense data (for example, the visual system tries to detect contours in order to identify entities). For effects on the quantum level to have an effect on the macro level of nerve cells and ultimately muscle cells, they would have to pass through those layers, starting from the physics of atomic forces, and moving up to the macro-world level of biological molecules, cells, cell structures, neural networks, etc.

The neurobiologist William H. Calvin dismisses this idea of a quantum mind (or "quantum consciousness")—the notion that our consciousness is somehow rooted in the physical phenomena of quantum mechanics—as a "janitor's dream" with the janitor of a high-rise building working in the basement (the quantum world), having to influence indirectly what is happening in the penthouse (the macro-world of biology).[1]

---

[1] cf. Calvin, 1996, p. 181.

The janitor's dream describes only one direction, namely how something like a quantum mind could *affect* the macro-world. For example, when we tell each other about our subjective experience of consciousness, information travels from our consciousness to our muscles, causing our hands to write and our vocal cords to move. We connect the information from our "mind's eye" that experiences consciousness to our language output.

But how a quantum mind could *perceive* the macro-world is yet another mystery altogether. After all, information also travels from our senses to our consciousness. The brain gathers sense data (for example, the question "Do you hear the noise of the train in the distance?"), the information is processed by the brain, perceived consciously, an answer is consciously decided on, and then the answer is transformed into sound waves using one's voice ("Not only do I now hear the train in the distance, I also know that I hear the train in the distance."). When you look at an apple, your eyes process the image and create a signal that activates various parts of your brain. These activations cause changes in the quantum state of individual atoms. A quantum mind would have to (somehow) "read" changes to produce a "quantum-level concept" of the apple. If the quantum mind then decides (how?) to reach for that apple, it would have to encode that decision back into microscopic quantum fluctuations that in their entirety produce a macroscopic pattern in the brain to activate the right set of neurons to cause the action of grabbing the apple. In summary, a quantum mind would:

1. First, "read" quantum level changes caused to neurons by the senses; then
2. Decipher this raw data back into high-level concepts (for example, "apple"); then
3. Make a decision on the quantum level about what to do with the apple; and finally
4. Encode that decision back into quantum level fluctuations to produce a macroscopic pattern.

But that is not a viable model for a quantum mind because a quantum mind would itself need a separate brain-like structure to translate the signals between the two systems. We already have a way of synchronizing multiple microscopic elements to produce a concerted change on a macroscopic level: that *is* the macroscopic world (the brain!) and there is no need to interpret and re-interpret data on a quantum level. That *is* what the brain itself is already doing. The proposed architecture of a quantum mind (see Figure 6.5) could be simplified by drawing an arrow from the left neurons box to the right neurons box, skipping the quantum mind altogether. Neurons would then be communicating with each other directly instead of having to hop back and forth between the macroscopic world and the quantum world.

In other words, a quantum mind could use the very brain it is reading to calculate what high-level information the brain is currently processing. Like the aforementioned janitor could go through each level of the building, pick up pieces of trash like a detective, and ask each tenant questions. By that standard, we could also argue that a bicycle wheel works using quantum mechanics because the molecules that make up the metal and rubber interact with each other on a quantum level. Of course, this raises the question: *Why do we even need to bring up anything related to quantum mechanics or an immaterial mind?* Because if the quantum brain uses the *material brain* to create consciousness, we can simply skip the idea of quantum mind and focus on how the brain works—just like we would focus on the wheel and chain when trying to understand how a bicycle works.

While there are exceptions to this idea of the quantum world not affecting the macro world (for example, the photoelectric effect), in this case the brain is more like a bicycle. If we cannot explain how one can ride a bicycle with its wheels, frame, and chain, we surely cannot explain it by referring to its molecular structure. Or even simpler: to grab a cup of coffee, you do not have to translate the cup and your hand into their molecular structures and calculate how they would interact with each other on that level. Most likely, evolution has found a way to optimize fundamental chemical and electrical processes (like neurons firing nerve impulses) in the brain at the subatomic scale. That does not necessarily mean that the brain has a "quantum mind." This is also in line with what we have learned about quantum computers in *Philosophy for Heroes: Continuum*: all they are doing is running controlled physical experiments, and then using the results to speed up certain calculations.

Just because the two concepts of quantum theory and consciousness have yet to be fully understood does not mean that they have anything to do with each other.

We can conclude that the value of the "janitor's dream" is very limited. Placing consciousness into the quantum world (as part of a "quantum mind") does not explain the observable macroscopic effects (muscle movements) of consciousness. We have learned, though, that the brain needs some kind of structure that is responsible for calculating what is going on at a macroscopic level—like a central information point at an airport gathers all the data about incoming and outgoing flights. The brain needs to know a summary of what it is thinking. We will address this issue again later in Chapter 6.3.4.

Given that we do not have any evidence for a separate quantum mind or for any communication from our biological brain to that supposed quantum world brain, we should consider that consciousness is a system consisting of macroscopic building blocks of reality (for example, the neurons and axons of the brain or interconnected brain parts). With the quantum mind hypothesis out of the way, we can continue to focus on other monist theories of consciousness. To arrive at such theories, we will first discuss in Chapter 6.2 what elements of the brain could be part of a system that produces consciousness.

6.2 The Search for Consciousness

If we dismiss the idea of a quantum mind and if we accept that consciousness does not reside in the immaterial world, then consciousness has to relate to one or several parts of our material brain. In this section, we will determine if there is a single brain part responsible for consciousness, if consciousness is an *emergent property* of the brain, or if several brain parts are involved in consciousness.

Most monist theories of consciousness imply that there is some kind of single "observer"-like element in the brain, even if the theories do not spell it out. This is what the philosopher and cognitive scientist Daniel Dennett calls the "Cartesian theater."

> Cartesian materialism is the view that there is a crucial finish line or boundary somewhere in the brain, marking a place where the order of arrival equals the order of "presentation" in experience because what happens there is what you are conscious of. [...] Many theorists would insist that they have explicitly rejected such an obviously bad idea. But [...] the persuasive imagery of the Cartesian theater keeps coming back to haunt us—laypeople and scientists alike—even after its ghostly dualism has been denounced and exorcised.

—Daniel Dennett, *Consciousness Explained*

CARTESIAN THEATER · The *Cartesian theater* is a term coined by Daniel Dennett to criticize most of the contemporary explanations of consciousness. At their core, such explanations all share the view that there is some sort of miniature person ("homunculus") or entity within the brain looking at what we are looking at—an idea which ends in an infinite series of subsequently ever-smaller Cartesian theaters with ever-smaller homunculi.

Instead of admitting that they are mind-brain dualist theories, most modern theories of consciousness stay away from discussing the Cartesian theater and only describe individual aspects of consciousness, without explaining the whole picture. Dennett criticizes scientists who look only at one part without looking at the whole picture. Each study pushes the responsibility to find the final "actual observer" to someone else further down the line.

The Cartesian theater fallacy demonstrates that there is no brain part or magical objective observer that takes over the brain and starts being a "brain in a brain." While the brain parts compete, such competition does not imply that they are conscious actors themselves. Because after all, who would control *that* brain or brain part? We could have another "brain" observing the previous brain and so on, ending up with the observer having in itself an observer observing the observer, and so on, ending up in an infinite series of observers. This is the so-called fallacy of the *Homunculus argument* (see Figure 6.7).

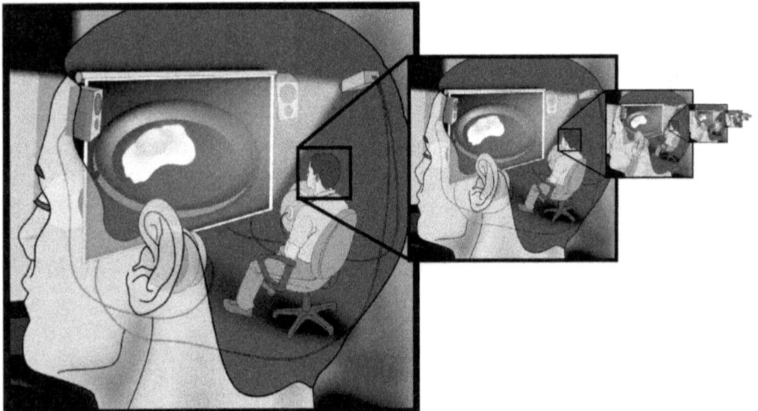

Figure 6.7: An illustrated example of the fallacy of the Homunculus argument: the observer himself is observed by an (even smaller) observer and so on (image source: Jennifer Garcia, wikimedia).

> **HOMUNCULUS ARGUMENT FALLACY** · The *homunculus argument* is the fallacy of trying to explain consciousness by another (smaller) conscious person (the "homunculus") observing and steering you. The problem with this explanation is that it remains unexplained how this smaller homunculus subsequently experiences consciousness.

This argument is reminiscent of the discussion in Chapter 6.1.4 about the role of the observer in quantum mechanics and consciousness. If the world exists only when we are observing it, who is observing us? This creates an infinite loop, too, with an undefined "original cause" (well, or "consciousness") at the end of the line, but does not really explain anything that is going on. With that in mind, it seems that consciousness is *both* finite and infinite.

In summary, the "Cartesian theater," in which there is some sort of miniature observer sitting in the brain and observing the incoming data, is not a satisfactory explanation of consciousness. Even if we had concrete proof of such an observer, we would simply move on and start examining *that* observer and question how *it* worked.

6.2.1 Emergent Property

> **EMERGENT PROPERTY** · An *emergent property* is a property of a system that emerges only when its parts are combined or interact with each other. Individual parts of that system do not have the emergent property themselves. For example, the division of labor in ant colonies allows the ants to be more efficient than if individual ants fended for themselves.

There is no central "evolution" or "creativity" authority directing nature to come up with novel solutions. We cannot point to a plant or animal and declare "that is evolution." Evolution is a process and the way lifeforms are selected is an emergent property of this process.

As such, it seems natural to ask whether consciousness is such an emergent property of our brain. Just like fish form schools that seem to behave like single entities, neurons could form consciousness.

One way to approach the question is to take a whole brain and then think about what would happen if we removed individual neurons.

As a comparison, imagine being a ruler of a country. Your cabinet advises you on policy decisions and you make the final call. Now people leave your cabinet one by one. You could still make decisions, although they would be less informed and may cause hardships among the population. In that regard, the brain is not much different: the sum of your brain's neurons is like a committee. Even losing neurons, you are still conscious and (your committee) capable of acting but your decisions and movements will be less refined.

While losing many neurons within a short period of time is usually lethal or leads to irreparable coma, if this damage happens slowly, you might not even notice it. For example, there was a case of a man whose brain mass was, over time, compressed to less than 10% of its original mass. While he had a reduced IQ of 75, he was still fully conscious.[2] The loss occurred due to a huge fluid-filled chamber that took up most of the room in the center of his brain over the course of 30 years, leaving little more than a thin sheet of actual brain tissue pressed against the skull. The loss of neurons happened globally, meaning that only volume—not connectivity or brain parts—was lost. In a way, it seems that the question about consciousness is like the question about the pile of sand we discussed in *Philosophy for Heroes: Knowledge* (Sorites paradox): how many grains of sand do you have to pile up in order to call it a "pile" of sand? One? Two? A thousand? If 90% of the brain can be removed, what would happen if even more were removed? Can a single neuron be conscious or create consciousness?

---

[2]Feuillet, Dufour, and Pelletier, 2007.

The limit between a conscious and non-conscious brain seems to be the structure, not the number of neurons. This implies that consciousness is probably not an emergent property of individual neurons but a product of how individual brain parts are connected to each other—a system we could clearly define and describe. Now, if we can find instances of localized damage in the brain that have resulted in impairments of consciousness, we can determine that those brain parts are elements of the system that creates consciousness. This is in line with how scientists have been able to discover the functions of the neocortex by correlating localized damage to specific cognitive impairments.

If there is no single "final" observer and if consciousness is not an emergent property, the remaining alternative is to look at consciousness as a system with many different brain parts interacting with each other. To examine this option, let us look at what we most prominently connect with consciousness, namely our sense of sight. To better understand the way our eyes send signals to the visual cortex, let us look at Figure 6.8 and examine what various types of damage to our visual system would do to our visual experience:

- Damage to 1 or 2: Damage to the eye or optical nerve can result in the loss of sight in one eye. The effect is the same as closing one eye.

- Damage to 3: Splitting the optic chiasm in the middle means that the left visual cortex no longer receives the right side of the visual field of the right eye, and the right visual cortex no longer receives the left side of the left visual field of the left eye.

- Damage to 4 or 5: Damage to the connection between the optic chiasm and the thalamus results in the left and right visual cortex receiving sense data only the left visual field or only the right visual field.

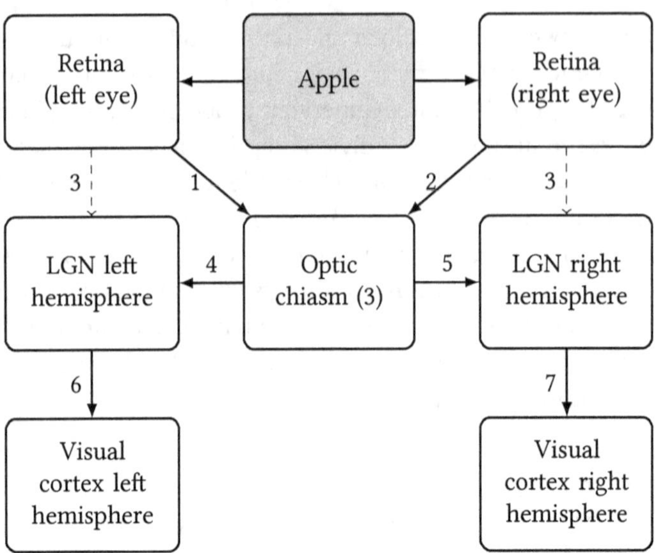

Figure 6.8: The basic process of perception, from initially seeing the object (in this case, an apple) to the pre-processed image in the cortex.

6.2.2 Blindsight

If the primary visual cortex itself (or the nerves leading up to it from the LGN, see 6 and 7 in the diagram) is damaged, this damage results in the same impact on the visual field as damage to 4 or 5. But the affected person can experience "blindsight" on the blind side of the visual field as some of the information is transferred from the thalamus to other parts of the brain.

> **BLINDSIGHT** · Someone suffering from *blindsight* reports that he cannot see. However, experiments show that he can react to visual cues. As a result of damage in the visual cortex, information from the retina arrives in the midbrain but does not undergo conscious processing through the visual cortex.

Sufferers of blindsight have perfectly working eyes, yet they *report* that they are blind. Given that their visual cortex never receives visual sense data, they cannot report what their eyes are seeing. But they can, for example, navigate through a room with obstacles much more effectively than blind people whose visual system is damaged *before* the LGN (their eyes or the optical nerves leading to the LGN). Even under test conditions[3] where they are presented visual stimuli, people suffering from blindsight scored significantly higher than random chance would suggest—at least when they are asked to rely on their instincts or "just guess." It is like asking someone without sight impairments to pass an obstacle course without thinking about the obstacles and instead only trusting her muscle memory and instincts (her "blindsight"). Bruce Lee put it succinctly:

> Don't think. *Feel.* It's like a finger pointing at the moon. Do not concentrate on the finger, or you will miss all of the heavenly glory.

> —Bruce Lee

The fact that people suffering from blindsight have the ability to accurately grasp for objects in their blind field tells us that there is already some processing of three-dimensional structures and movements in the LGN, providing the other brain parts basic information about the environment.

Theories regarding blindsight include the assumption that some data (around 10%) from the LGN travels not just to the primary visual cortex at the very beginning of the ventral and dorsal streams but also further down the line to other brain areas (for example, the *visual association area* that is adjacent to the primary visual cortex).

---

[3] Stoerig and Cowey, 1997.

Blindsight contradicts our intuitive understanding of blindness. If we cannot report what we see, how can our brain still see and "report" its findings to us with intuition? It certainly seems an uncomfortable thought, not being in the driver's seat, especially while our usual inner experience suggests that we are in control. For us to be conscious of something, that information needs to be processed by particular brain parts. If it is not, it can still affect us, but we cannot translate it directly into words. The underlying principle is actually common in daily life. Try to explain to someone how to keep balance on a bicycle and you discover that you can do something (ride a bike) which you cannot explain with words. That is the reason professional trainers have to learn their movements in an abstract way, too, in order to correctly verbalize them to their students—even experts do not have direct access to their own bodies' optimized programs of their cerebellum.

It seems that something can catch the attention of our brain, but we do not necessarily become *aware* of it. In this context, awareness means that the information is available for conscious processing. For example, your mind might wander when you are on the phone and taking a walk. It might take a while before you become again aware of the fact that you are walking. If there are no obstacles in your path and the conversation is deep enough, your brain might still be processing walking along the path, but it never reaches your awareness. This shows that it is possible to have attention without awareness.[4]

> **AWARENESS** · *Awareness* is a description of the process of attention. Something can grab the attention of your brain, but to talk about it, you need a model of what is happening in your brain. Awareness is such a model.

[4]Webb, Kean, and Graziano, 2016.

6.2.3 Split-Brain Syndrome

When the connective tissue between brain regions is damaged, this has implications for many regions of the brain. By far the biggest connective bridge between brain parts is the *corpus callosum*, connecting the left and right hemispheres of the brain (see Figure 6.9). To understand the role of the corpus callosum, we need to understand the role of each hemisphere.

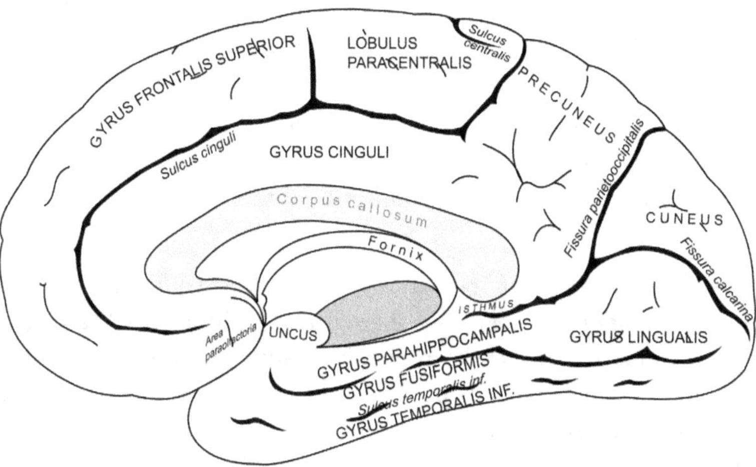

Figure 6.9: The location of the corpus callosum between the left and right hemispheres (image source: Wikipedia).

CORPUS CALLOSUM · The *corpus callosum* connects the left and right brain hemispheres, coordinating tasks requiring both sides.

In healthy people, the most obvious difference between our left and right hemisphere shows up in our handedness. Not all people are right-handed—some people are left-handed, and a few are ambidextrous. For simplicity, we will focus on right-handedness for now. As our right hand is controlled by the left hemisphere and our left hand by our right hemisphere, we can see a clear division of labor between both hemispheres (so-called *lateralization*); for people with right-handedness, there seems to be a preference for fine-grained control in the left hemisphere. This relates to what we have learned in Chapter 5.4.1 about the involvement of the left supramarginal gyrus in tactile sensory data and tool use. Still, even with this division of labor, the two hemispheres have to work together. When you look straight ahead and then try to reach for an object in your left visual field with your right hand, information from your right visual cortex has to be transferred to your left hemisphere to coordinate the hand movement with visual information. This information transfer is made possible by the corpus callosum that connects both hemispheres.

However, why is the architecture of the body so confusing, with the left hemisphere controlling the right side of the body and the right hemisphere controlling the left side of the body?

The main reason is that evolution looks for short-term advantages. This "entanglement" between the left and right side is thought to have originated very early in our evolution. The theory is that at one point in time, the body plan (a "blueprint" detailing aspects such as symmetry, segmentation, and limb disposition) of an early vertebrate was twisted by 180 degrees which left this so-called decussation in its wake.[5]

Beyond how the right hemisphere controls the left hand and the left hemisphere controls the right hand, the hemispheres are also specialized in different tasks (see Figure 6.10). As a rule of thumb regarding the architecture of both hemispheres, the left hemisphere of the brain usually deals with the concrete, while the right hemisphere deals with intentions, interpretation, and hidden meanings.

---

[5]Lussanet and Osse, 2012.

Left hemisphere	Right hemisphere
Right-hand control	Left-hand control
Right field vision	Left field vision
Language	3D forms
Concepts	Measurements
Math	Face recognition

Figure 6.10: A (non-comprehensive) comparison of the different specializations (the lateralization) of the left and right hemispheres of a human's neocortex.

The advantage of this architecture is that we can use it to learn more about the functioning of the brain. If the corpus callosum gets damaged, we can draw different conclusions about consciousness and its involved brain parts depending on what symptoms people experience in such a case. If people with a damaged corpus callosum retain normal consciousness, this might be an indication for mind-brain dualism. If they lose consciousness, it would mean that both hemispheres are necessary for a conscious experience. If we find that by splitting up the brain, we create two separate consciousnesses, this would contradict our previous finding that consciousness is probably not an emergent property. If we create two separate consciousnesses that have limitations, consciousness is either an emergent property, or identical structures exist in both hemispheres.

When the corpus callosum is damaged, people can experience *split-brain syndrome*. In the past, such damage was caused by surgery to prevent life-threatening injuries caused by severe epileptic seizures. Disconnecting the left and right hemisphere prevented the epileptic seizure from spreading to or oscillating between both hemispheres. Nowadays, this type of surgery is applied only in very rare cases, given its significant adverse effects. A severed corpus callosum means that the left hemisphere cannot directly access the right hemisphere and vice versa.

> **SPLIT-BRAIN SYNDROME** · The *split-brain syndrome* can occur when the *corpus callosum* is damaged. This leads to problems with communication between the left and right brain hemispheres and complicates some tasks requiring both sides.

Some tasks might be more difficult for the split-brain patient to solve, as he needs to use different paths to connect both hemispheres. Because people with split-brain syndrome can compensate for the lack of connection between hemispheres, it is not easy to identify those who have the condition. Doing so requires specific experiments to determine whether or not the connection is missing. One experiment commonly used for split-brain patients involves instructing the patient to stare at a dot on a computer screen and then showing her different words and pictures on the left and right side of the dot. When asked what she is seeing, she is unable to report on what she sees on the left side of the dot (see Figure 6.11). This is because her language center is in the left hemisphere. Without a corpus callosum, it has no direct access to the right hemisphere which processes the left visual field. And without connection through the corpus callosum, this information from the right hemisphere is not available in the left hemisphere.

When the patient is asked to close her eyes and draw with her *left hand* what she has seen on the left side, she is able to do so (see Figure 6.12). This shows that in patients with a severed corpus callosum, the information arrives and is correctly processed in both hemispheres, but each hemisphere is unable to access the information from the other hemisphere.

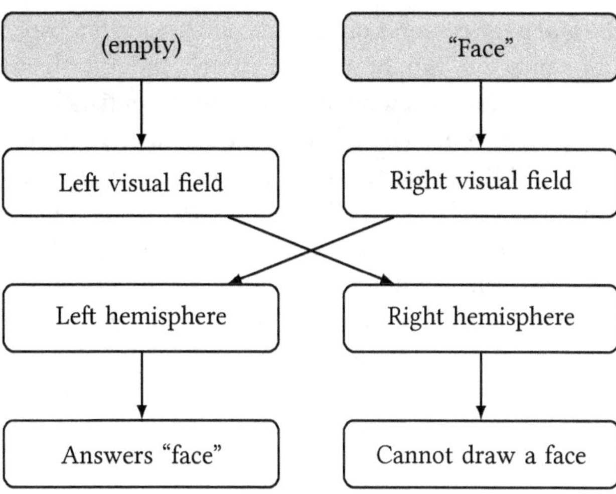

Figure 6.11: The split-brain patient can tell you what is on the left side but (due to the missing connection between the left and right hemisphere) cannot draw it with her left hand.

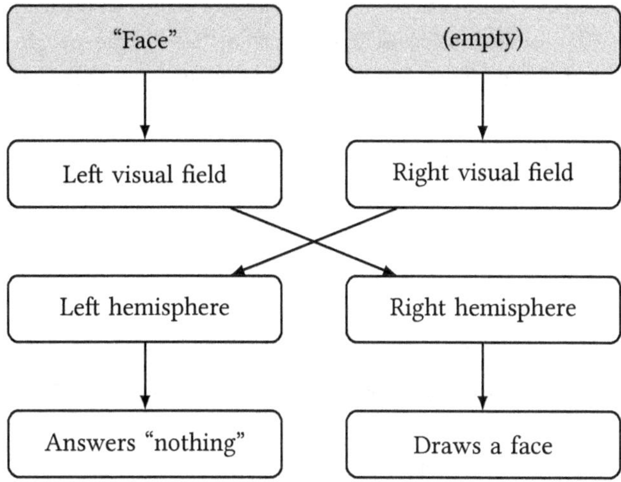

Figure 6.12: The split-brain patient can draw with her left hand what is on the right side. But due to the missing connection between the left and right hemisphere, the patient cannot tell you what is on her right side.

While the language center is often located in the left hemisphere and the right hemisphere cannot produce speech, the right hemisphere can read and understand simple questions. Scientists have found ways to train the (otherwise silent) right hemisphere in split-brain patients to point the left arm (which is controlled by the right hemisphere) to, for example, either "Yes" or "No" written on a blackboard. That way, the researchers were able to pose questions to both the left and the right hemisphere of a split-brain patient—with surprising results. The (silent) right hemisphere could correctly answer a variety of questions (where the subject is, his or her gender, etc.) but for some questions, the answer differed from the left hemisphere's response. For example, one patient's left hemisphere told the scientists he believes in God, while the right said he does not. On closer inspection, this is not surprising, though, as each hemisphere bases its decisions on different data and, without the corpus callosum, the two sides cannot negotiate the final decision.

> Imagine our surprise when we noticed that in patient LB the left hemisphere said it believed in God whereas the right hemisphere signaled that it was an atheist. The inter-trial consistency of this needs to be verified but at the very least it shows that the two hemispheres can simultaneously hold contradictory views on God.

> —Vilaynur Ramachandran, *The Emerging Mind*

We see similarities between the brain of a split-brain patient and the nervous system of an octopus. Some two-thirds of its neurons are located in its limbs instead of its central brain. They can move independently of each other, even after being severed from the main body. Both the split-brain syndrome as well as examples like the octopus lead us to question the intuitive idea of a simple one-to-one relationship between "number of brains" and "consciousnesses."

Our observation actually points to split-brain patients having *two* consciousnesses, each being somehow partly conscious. Both sides seem to be "conscious" but neither side has conscious access to the information or abilities (like spoken language or comprehending three-dimensional forms) of the other hemisphere. For example, as discussed above, the right hemisphere of a split-brain patient has no access to the language center in the left hemisphere and thus cannot verbally explain how it experiences consciousness. While we have uncovered more about consciousness with this examination, we need to dig deeper. From a certain perspective, the split-brain syndrome could be compared with psychophysical parallelism (see Chapter 6.1.3). There, mind and brain originate from the same substance but were split at some point in time, running in parallel from then on. Likewise, in patients with split-brain syndrome, the brain was whole at some point, but both brain hemispheres had to run in parallel after the corpus callosum was severed.

6.2.4 Hemispatial Neglect

Can we see something but not be aware of it?

For sufferers of *hemispatial neglect*, the ability to report on one side of their visual field is impaired. However, hemispatial neglect is not merely a problem of a subject's vision. Instead, the issue is that any *conscious experience* of this part of the field of vision is impaired while the sense of sight (the eyes and the visual cortex) is working perfectly well.[6] Left-sided hemispatial neglect is more common than right-sided hemispatial neglect. This is because attention to the left side is processed only in the rTPJ, while attention to the right side is processed in both hemispheres.

---

[6] Parton, Malhotra, and M Husain, 2004.

> **HEMISPATIAL NEGLECT** · Someone suffering from *hemispatial neglect* lacks consciousness of half of his visual field. The person is not aware that his vision is impaired in any way, making the condition different from blindness in one eye. People with this condition have to learn abstract strategies as a way of coping.

The important distinction between an impaired field of vision and the condition of hemispatial neglect is that sufferers of hemispatial neglect do not recognize that anything is missing in their visual field. The challenge for people with hemispatial neglect is not only that they are *not conscious* of something, but also that *they do not realize that they are not conscious of it.* For example, when asked to copy a clockface, a patient with hemispatial neglect ends up drawing a circle with the numbers 1 to 6, omitting 7 to 12 on the left side (or drawing all numbers 1 to 12 on the right side). If he was just *visually* blind on the left side (hemianopia), he would be conscious of the fact that he could not see the left side of the original clockface and turn his head.

In one study, subjects who were thought to have this condition were presented simultaneously with two line drawings of a house. In one of the drawings, the left side of the house was on fire and the right side was fine. Patients with left-sided neglect (and right-sided brain damage) reported not seeing anything wrong with the house that was, in the picture, on fire. They stated that the drawings were identical; yet when asked to select which house they would prefer to live in, they reliably chose the house that was not burning.[7]

One explanation is that each hemisphere can still "communicate" with the other hemisphere indirectly through lower brain functions. For example, the left and right parts of the hippocampus are connected, and so are the left and right parts of the amygdala. Likewise, the hormonal system acts on both hemispheres. In that regard, people with hemispatial neglect can be influenced without being con-

---

[7] Marshall and Halligan, 1988.

scious of that influence. A patient with hemispatial neglect "just knows" and tries making up some rationalization to justify her feelings in order to appear rational. She does not have conscious access to the image of fire on the left side of a house, so in her verbal explanation of why she feels uneasy when looking at the picture, she will use some magical or far-fetched explanation of why she thinks something is wrong with it. Another example would be ducking from an approaching object. Having a ball thrown at her from the left side might cause her to duck, but she might be not conscious that there was a ball or why she ducked. She has access to the emotional evaluation of a situation but not the abstract reasoning behind it. This result can be interpreted as: in the affected patients, there is a disconnect between what their brains process and the information they have access to when trying to rationalize their decisions.

To what could we compare such an experience?

Let us say you are driving in your car and the car's tail light is broken. You would still accelerate and break as you usually do, not conscious of possible danger. You might notice that something is wrong because other drivers flash their lights at you, but you have no idea why. Similarly, a sufferer of hemispatial neglect is not conscious that anything is missing on one side of her visual field. She would put make-up on only half of her face, comb only one side of her hair, and walk out of the door as if nothing were wrong—just like you would enter and leave your car like everything was fine. She would not be aware of what is wrong when encountering people reacting strangely to her behavior.

The difference compared to driving with a broken tail light is that even if she consciously knows that she suffers from hemispatial neglect on, for example, her left side, she is still unable to do something about it in a direct way. She is only conscious of her lack of "left side" in an abstract way because that was her diagnosis. She has no option to turn her head to be able to see what is missing from her vision as she is not conscious of her limited visual field. Without therapy, hemispatial neglect can even lead to a loss of attention to the limbs on the left side. She has to use her prefrontal cortex to develop habits that continually demonstrate to the rest of her brain that she indeed still has a left arm. To cope with their impairment, people with hemispatial neglect *cannot* rely on a separate "self" sitting in some sort of driver's seat of their mind or body. To overcome cognitive defects like neglect or delusions, they have to learn strategies to cope. The major challenge with cognitive defects is that the people who have them are not necessarily conscious of the fact that there is a problem.

To imagine how it feels to explain something to which you do not have conscious access, try to explain why your favorite color is your favorite color. This might strike you as insulting, but your story will probably be an invention to justify your preferences (e.g., "I like blue because it's the color of the sky"). As long as you do not know your neural configuration in your visual cortex but try to come up with a reason through introspection alone, you will end up confabulating instead of giving actual insight into how your brain works. This is because your left hemisphere, which provides verbal answers, believes that its interpretations and conclusions are correct. The actual cause might be some neural preference, while your rationalization for your preference is probably a non-rational explanation pointing to an unspecified "hunch" or "feeling" about why one color is preferred over another.

Some of our preferences can be changed. We can put ourselves into situations to train particular skills and form new habits. This way, the prefrontal cortex can change the rest of the brain in an indirect way. For example, by climbing up a diving tower and jumping into the water, the prefrontal cortex can demonstrate to the rest of the brain that you are able to accomplish this activity and survive. Similarly, going to the gym might initially require a lot of overcoming. After a few weeks, it becomes a habit and part of your daily routine.

What can we deduce from our examination of the split-brain syndrome, hemispatial neglect, blindsight, and AHS as these relate to consciousness?

We see that we cannot be aware of what is not fed to us by the brain. For example, as experiments with patients suffering from hemispatial neglect show, parts of the visual field we are not aware of are not "blacked out." We cannot simply turn our head to cover the entire visual field. Instead, it seems that our options to react to our environment are limited by the attention mechanisms provided by our brain. There is no homunculus sitting in our brain with an objective overview of incoming sense data.

This leads us to the question what we mean when referring to ourselves as "us." As experiments with people suffering from blindsight or hemispatial neglect show, sense data is still processed in the brain, but it is not directly available in our consciousness. We can still act on a subconscious hunch or feeling, but there is a qualitative difference between making a decision based on a hunch and being able to explain why you are making the decision. With obstacles in the way, it is certainly sufficient to "just know" your way around them (as people with blindsight do). Having the information available in your consciousness could help you to plan a path through the room of obstacles or think about a strategy for how to remove the obstacles altogether. In that regard, hemispatial neglect is the opposite of someone suffering from a monothematic delusion (see Chapter 5.3.6) who is still fully aware of his environment, but who is unable to properly connect emotions with what he sees.

Consciousness seems to be neither an emergent property from neurons, nor the product of a single brain part. We have confirmed this by looking at damage to the visual system, the connection between the hemispheres, and syndromes that affect awareness but not attention. With this knowledge, we will examine what specific brain parts are connected with each other to form a *loop of consciousness*.

6.3 The Loop of Consciousness

If consciousness were an emergent property of the brain, adding neurons would make you more conscious, and removing neurons would make you less conscious. While we have looked at examples of impaired consciousness, they were related not to the extent but to the *location* of the damage. This points to consciousness being the product of a system within the brain rather than simply an emergent property from the sheer number of neurons in the brain. If this assumption is true, we can formulate a *mechanistic theory* of consciousness and (at least on paper) *build* a conscious system. In this section, we will examine what happens to the sense data after processing in the occipital, parietal, and temporal lobes. Then, we will put the process of consciousness into a diagram.

> **MECHANISTIC THEORY** · A *mechanistic theory* is a theory that explains a system as if it were a computer program or a mechanical machine with gearwheels: you can understand it by tracing the input to the output. Pedagogically, a mechanistic theory is of great value as we can explain the essential workings of a system step by step using a diagram with boxes and arrows.

Let us reexamine the idea of the Cartesian theater from Chapter 6.2. Instead of arranging the homunculi in an infinite series, we put a finite number of observers into a *loop* observing each other. Just as you can create a (theoretically near-infinite) number of mirror images by placing two mirrors nearly opposite of each other, we could be somehow fooled by a very high number of loops of consciousness. The mirror images do not exist; they are part of the process of reflecting light back and forth between mirrors. Similarly, we should look at consciousness as a looping system of information (instead of a fixed entity) being perceived, processed, expressed, and perceived again.

> **Did you know?**
>
> This process of going back and forth is actually how we form knowledge. By reflecting upon our perceptions, integrating them, and then using that newly found knowledge to reflect upon our own cognition, we get a better and better idea of how the world works. In other words, whenever we learn something new about the world ("what is?"—ontology), we have to reconsider how we learn facts about the world ("how do we know?"—epistemology), and then, with our new way of interpreting what we perceive, again reconsider *what* we know about the world, and so on.
>
> $\longrightarrow$ Read more in *Philosophy for Heroes: Knowledge*

One candidate for such a loop can be found in the thalamus. The thalamus is not just a relay station with some pre-processing going on. For example, the visual cortex uses the LGN to focus on particular aspects of an image and enhance it. While the LGN sits on the connection between the retinas and the visual cortex, it actually receives 95% of its input from sources *other* than the eyes, that is:

- The superior colliculus, which is responsible for visual tracking and eye movements;

- The visual cortex which creates a feedback loop with the LGN; and

- The midbrain and brainstem which are responsible for pupillary light reflexes, smooth pursuit of moving objects, the accommodation reflex, antinociception, REM sleep, arousal and sleep-wake cycle, attention and memory, behavioral flexibility, behavioral inhibition and psychological stress, cognitive control, emotions, neuroplasticity, posture, balance, and preparation for fight and flight.

This shows that the LGN pre-processes everything coming from the eyes and is in a feedback loop with the visual cortex. Sense data does not just flow in one direction—bottom-up from the eyes to the visual cortex—there is also information that was already processed in the visual cortex flowing back into the LGN (see Chapter 5.3.1). In both cases, the neocortex provides high-level information to adjust the filters in the eyes and nose. This is similar to the olfactory system that we have discussed in Chapter 5.1.3. There, the neocortex also "primes" the olfactory system's cells to increase its sensitivity to particular smells.

While those two examples seem promising, the problem is that each loop degrades the sense data. Like the telephone game where people sit in a circle and whisper the word they (think) they heard from their neighbor, the looping image (or smell) information will no longer resemble the original image captured in our retina (or the original smell captured in our nose).

Still, even without degradation of the signal, we are left with the questions where *exactly* something like consciousness would arise in this loop, and why we claim to experience it. It is magical thinking to assume that just because something is (potentially) infinite, looping very often, very small, or very large, new properties arise that have nothing to do with the process. For example, when crystals are cooled down, slower, more *complicated* (but not more complex) structures emerge (see the table with the snow crystals in *Philosophy for Heroes: Continuum*). It is just the same rules of geometry and molecular forces applied more often. Similarly, a lot of water droplets are moving when it is raining (complex), but the underlying concept of water falling from the sky is not difficult to understand (not complicated). In summary, just describing a "loop of consciousness" is not enough, we need a precise definition of *consciousness*. After all, a loop is just that, a loop. Where would the consciousness that we experience come from?

6.3.1 The Process of Consciousness

To map consciousness into a diagram, we start with Figure 6.13. At its core, there are two loops. The first loop simply goes through perception, consciousness, effects on our environment, and back to perception. We observe our own actions (and their effect on our environment) and use this information to think about our next action. Instead of reacting immediately to a new perception, we might first loop "within" our consciousness. This could be the activation of memories, or the mental planning of our next steps.

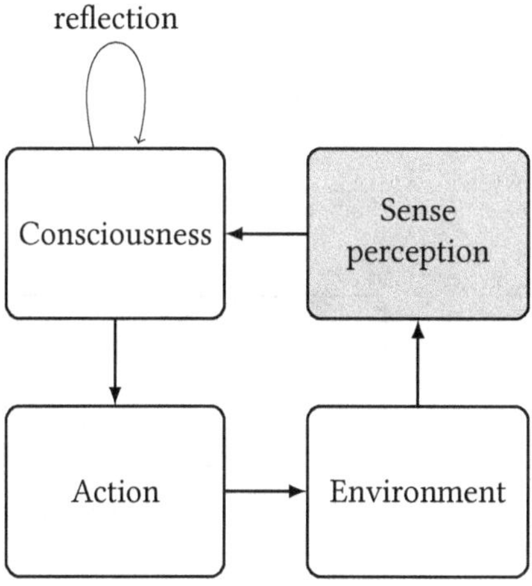

Figure 6.13: The basic process of consciousness contains two loops, one internal (via reflection), one external (via interaction with the environment).

Now, our job is to refine this diagram until we have replaced all occurrences of "consciousness" with actual brain parts. For the perception part, we have cleared up in Chapter 5.3 how information arrives in the visual cortex through the eyes and is then processed in the parietal lobe (identifying the "where") and temporal lobe (identifying the "what"). This process (see Figure 6.14) is as follows:

1. The apple reflects light rays which hit the retina of our eyes.
2. The *optic chiasm* combines the nerve signals from the retina into the left and right visual fields.
3. The *thalamus* pre-processes and then recombines the image data with possibly other sensory data.
4. The *visual cortex* receives the visual information.
5. The *parietal cortex* ("where") and *temporal cortex* ("what") analyzes the data.

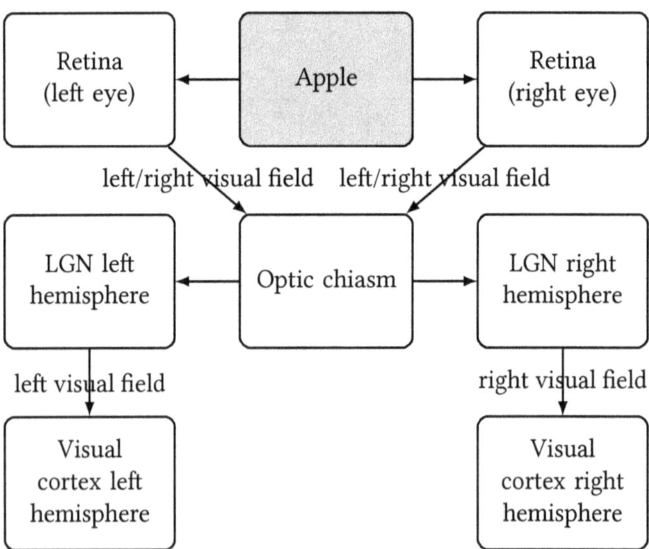

Figure 6.14: The basic process of perception from the object (in this case, an apple) to the pre-processed image in the cortex.

6.3.2 Working Memory

How do we access the brain's sense data?

Memory addresses the issue that sense data processed by the brain is fleeting. As discussed in Chapter 5.1.7, a basic approach nature took to remember previously perceived sense data was to release stress hormones, preparing our body for possible injury, and also to take one of the four actions: fight (proceed), flight (run away), freeze (try to not be seen), or fawn (try to please—more applicable in confrontation with other members of the same species). The chemicals will last for a longer time in the body, long enough to deal with the situation at hand—a kind of short-term memory.

With the analogy of neural committees, memory can be thought as every member of the committees being put into one of those four emotional states. The committees will still not remember, for example, a tiger, but they continue to stay in a fearful state until the hormones have subsided. Besides its function as memory, the hormonal system can also act like a prefrontal cortex and suppress other signals. For example, when running from a tiger, it would be good to suppress pain and thus you can keep going despite an injured leg.

In computers, so-called "regenerative capacitor memory" were used as a memory system. It was based on electrical properties of capacitors which you can "load" with a single bit (load or no load) and which keep that information for a while. To prevent the degradation of the memory caused by leaking of energy over time, the capacitor is regularly refreshed with its own bit-value. In the brain, we find a similar mechanism: to keep a thought pattern active, the solution is to continuously "read" and then immediately "write" the pattern back again, forming a loop and providing memory that can be accessed at any time. You can even create such a loop yourself to extend your short-term memory: simply tell yourself aloud the thing you want to remember again and again to keep it as an active thought pattern in your brain. The downside of this kind of memory is that, just like with the previously mentioned telephone game, the signal degrades with each loop.

In the brain, we have a variety of distinct memory systems based on loops that refresh themselves and thus allow retaining information for a longer amount of time:

PHONOLOGICAL LOOP · The *phonological loop* is an acoustic memory system of the brain that holds spoken words or sounds. It involves (among other brain parts, see Buchsbaum and D'Esposito, 2008) Broca's area and Wernicke's area.

VISO-SPATIAL SCRATCHPAD · The *visuo-spatial scratchpad* is responsible for temporarily storing visual and spatial information (involving the occipital lobes).

EPISODIC BUFFER · The *episodic buffer* is responsible for remembering the sequence of events, including the last state of an entity.

That these types of memory are separate can easily be demonstrated by trying to remember a phone number. By imagining one part of the phone number visually (using your visuo-spatial scratchpad), and the other part of the phone number as individual numbers repeating them with your inner voice (using the phonological loop), you can remember phone numbers more easily, at least in the short-term. Together, all those types of (short-term) memory form the so-called "working memory."[8] This is also the reason older phone numbers contained letters for better memorization, using the printed letters below each number on the telephone.

WORKING MEMORY · *Working memory* is the collection of a number of different (limited) short-term memory systems in the brain. The prefrontal cortex has access to the working memory.

---

[8]Chai, Abd Hamid, and Abdullah, 2018.

Adding working memory to our diagram results in Figure 6.15:

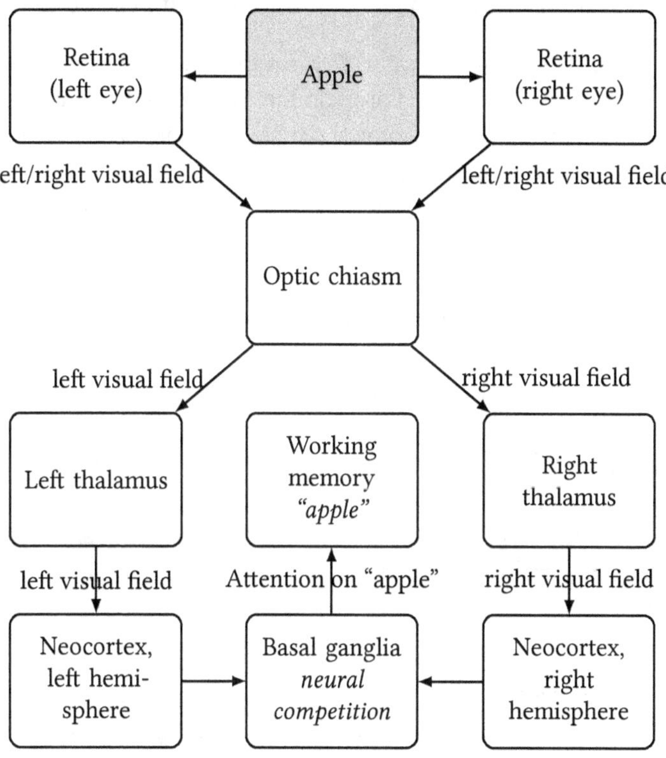

Figure 6.15: The basic process of perception from the sense data (in this case, the image of an apple) to the pre-processed image in the cortex.

We also have what is called *declarative memory* which connects long-term memories of sense data. We can find it in the amygdala (mapping thalamic sense input to emotions), and the hippocampus (connecting locations with each other for orientation and pathfinding). The latter has been shown by experiments with rats where specific neurons in the hippocampus became active when the rat was in certain locations.[9] The hippocampus is also involved in creating long-term memories in the neocortex. An example would be learning associations, that is, to connect two concepts with each other.

> **DECLARATIVE MEMORY** · *Declarative memory* connects one memory with another. For example, the amygdala maps thalamic sense input to emotions; the hippocampus maps places with each other for orientation; and the neocortex maps concepts with other concepts (a tree is a plant).

Declarative memory is processed by and temporarily stored in the hippocampus. Mainly during sleep, it is then copied for long-term storage into the temporal lobe and other places. Someone with damage to his hippocampus has a greatly reduced ability to form new declarative long-term memories. The affected person tends to forget what he did 15 minutes ago, and may repeat what he did just before.[10] But he can still use his working memory as usual and form new *non-declarative* memories. For example, he can learn new motor programs in the basal ganglia and cerebellum, or learn to recognize new images.

As concepts are knowledge about the world and seldom subject to change, they are stored differently than working memory. Instead of continuously looping thought patterns through the brain, neurons are trained repeatedly to adapt their configuration during REM sleep (dreaming). This can be imagined like a stream of water eventually digging a river bed into the ground: the more often a particular memory is replayed, the stronger the impact on the involved neurons. If a particular memory is connected to a strong emotional experience, this effect is magnified even more. Similarly, implicit memories like muscle memory are encoded in the configuration of individual neurons (in the basal ganglia and cerebellum), too, and the strength of the memories increases with repetition of the motor program.

---

[9] Skaggs and McNaughton, 1998.
[10] Squire, 2009.

While declarative memory is passive, external stimuli can activate a cascade where activating one memory leads to the next and so on. This applies especially to odors because of the direct connection of our olfactory system to the neocortex (and thus to memories). Consciously recalling memories works the same way: instead of recalling memories directly, we have to work through our memories step by step and find things we associate with a memory we want to recall. For this, the prefrontal cortex sends signals to the hippocampus to simulate sense data that would normally lead us to remember a particular thing.[11] The hippocampus then activates the neurons related to sense data in the temporal cortex. In that regard, the same pathway is used whether you see something that leads you to remember something (going from the thalamus to the hippocampus to the temporal cortex) or imagine something (simulating the sense data with the help of the prefrontal cortex, going from the prefrontal cortex to the hippocampus). You could also use imagination (see Chapter 6.3.5) to activate your visual cortex, working memory, thalamus, and ultimately your hippocampus. For example, when you are wondering where you have left your keys, you cannot ask your memory to magically provide you with the location (assuming you do not have a special place where you always place your keys). You have to trace your steps and ask yourself where, for example, the keys were when you were in the kitchen. By imagining the kitchen table, maybe you remember the jacket you have left in the kitchen, and that a jacket pocket contains the keys.

One takeaway is that memories are not movies we can simply replay but always scenarios we have to reconstruct piece by piece. In fact, we use the same brain parts to imagine a future scenario as we are using to recall a past memory.[12]

---

[11]Minxha et al., 2020.
[12]Schacter et al., 2012.

6.3.3 Stream of Consciousness

What happens to the sense data after it is processed in the occipital lobe, parietal lobe, and temporal lobe?

The TPJ (see Chapter 5.5) combines the ventral ("what") and dorsal ("where") streams to construct information about the subject ("Who is perceiving this?" "Whose thoughts are these?"). This data is then grouped together with the rest of the information from the parietal and temporal lobes into the working memory for high-level processing. For example, let us imagine a scenario of two people (Anna and Peter) in a room with an apple on a table. Peter is looking at the apple, and Anna is looking at Peter and the apple. Also, Anna is wearing a hat and is in love with Peter. Then the contents of Anna's working memory would be as listed in Figure 6.16.

Parietal lobe	Temporal lobe	TPJ	Working memory
A to B	eyes, apple, Peter	Peter	Peter is looking at the apple.
A on B	head, hat	Anna	Anna is wearing a hat.
A to B	eyes, Peter, apple	Anna, Peter	Anna is looking at Peter and the apple.
A to B	love, Peter	Anna, Peter	Anna loves Peter.
A	Anna	Anna	I am Anna.

Figure 6.16: Examples of the contents of the working memory based on input from the parietal lobe, the temporal lobe, and the TPJ.

Whether you ask another person, or another person asks you, in both cases, your brain hears the question "What do you see?" and your brain has to determine who is asking whom. Similarly, when answering the question, there might be multiple entries in your working memory about what is seen by whom. There is not only a list of items you see, but also a list of items the other person might see. When looking into someone's eyes and that person is looking into yours, your working memory has (at least) two entries: "I see you" and "You see me." Hence, each sentence contains not only raw sense data, but also information about *who* is perceiving the sense data.

The point is that by looking at Peter, Anna knows that he is, for example, looking at an apple. She has no direct access to his sense data but infers this information from the direction of his gaze. While the subjective experience between this abstract inferred information ("Peter is looking at an apple") and her own sense data (looking at Peter and the apple) is fundamentally different, this does not matter on the level of abstraction of the working memory. There, it is reduced to simply "Peter is looking at an apple" and "Anna is looking at Peter and an apple."

Without this additional information, the prefrontal cortex would not know what the real-world equivalent of the word "Anna" is. Likewise, without processing by the TPJ, Anna would not know who is looking at or loving whom. She would be unaware that Peter is looking at the apple, that she herself is wearing a hat, that she is looking at Peter and the apple, or even that she loves Peter. She might still *act* positively toward Peter, but she could not use that information for any higher processing like language, planning for the next day, or thinking about the social appropriateness of her actions. If the information is not completely missing but miscalculated or misattributed by her TPJ, she might end up (maybe falsely) thinking that her own emotions (also) belong to Peter ("Peter loves Anna").

If the brain does not correctly attribute sense data as being perceived by the self, it is not put into one's working memory (or it is attributed to someone else). This situation is similar to when something does not catch our interest in the temporal or parietal lobe: the corresponding sense data is filtered out. For example, in the "Where's Waldo?" children's puzzle book series, the reader is challenged to find a character named Waldo hidden in a group of dozens of people. The entire picture is available to the eyes but most of the details are filtered out by the brain. Only by scanning one part of the image after the next, Waldo can be found.

It could also lead to false memories: someone telling us about an event might turn into what feels like a first-hand experience in our memory. Likewise, this could be the origin of ghost stories or out-of-body experiences: attributing a voice or appearance to something outside the body instead of to oneself can cause serious emotional distress. The fact that we need an attribution of self to sense data also explains how short-term memory impairments can stem from TPJ damage rather than damage to parts of the brain that actually store working memory.[13]

This idea of a third stream besides the ventral and dorsal stream also helps us to better understand how AHS (see Chapter 5.4.3) can develop from damage to the rTPJ (or to connections from or to the rTPJ). The brain still receives visual and sense information from the left arm, but that information is never processed in the rTPJ for integration. The working memory will contain information that there is an arm, and that it is moving, but not information about *whose* arm it is. Eventually, the brain might conclude that it is not your arm. Any action initiated by the parietal lobe (utilization behavior) will then no longer be checked by the prefrontal cortex. The arm might end up acting out socially unacceptable or even self-harming behavior.

---

[13]Ravizza et al., 2011.

While the rTPJ is needed to recognize the difference between self and others and to properly attribute sense data to entities, the lTPJ translates this information into something that can be used by other parts of the brain. Remembering what we have discussed about the difference between the left and right hemispheres in Chapter 6.2.3, the left side is responsible for abstract language while the right side is about spatial relationships. To be able to speak about an object you are seeing and to use this abstract information to plan ahead, the lTPJ needs to be involved.

Another way to understand this is to look at the lTPJ as being responsible for concepts, while the rTPJ is responsible for measurements. You might be aware (with the lTPJ) that Peter, an apple, and the process of looking are involved. But you need the rTPJ to understand that Peter is *another* person, *which person* Peter in your field of view is, *where* the apple lies *in relation to* Peter, and that Peter is looking *at* the apple.

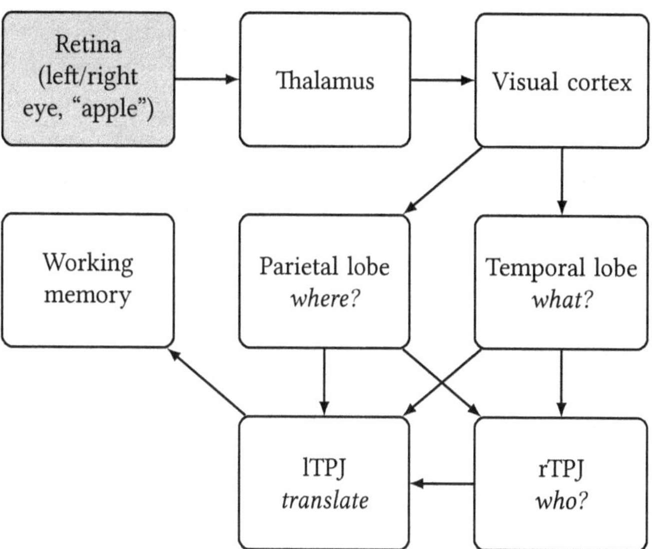

Figure 6.17: The basic process of perception from the object (in this case, an apple) to the pre-processed image in the cortex.

6.3.4 The Executive Functions

Having explained how sense data arrives in the working memory on the one side, we can narrow down the definition of consciousness by examining the other side, namely the last steps leading up to a physical action. If it is not a reflex or other non-conscious action (breathing, swallowing, coughing, etc.), the primary motor cortex generates motor programs to achieve a goal. The basal ganglia select a single motor program and coordinate the final movement program with the cerebellum. The selection process relates to the basic theory of how we make decisions that we have explained in *Philosophy for Heroes: Continuum*: competing neural committees refine the motor program in several steps and the "best" program is selected. If the motor program is new, this process can take a moment (or could take multiple physical attempts if the movement is complex, like a child trying to walk for the first time).

The executive functions are provided by the prefrontal cortex. Instead of initiating an action, the main task of the prefrontal cortex core function is to suppress whatever catches your attention if that action does not align with your long-term plans or personal preferences. For example, not eating the candy bar in front of you to fulfill your long-term goal of being healthy requires prefrontal cortex activity. Additionally, instead of suppressing a single action, the prefrontal cortex can also suppress every activity but one. This can help with focusing on one particular activity. But instead of trying to directly suppress most neural activity in the primary motor cortex, the actual suppression happens in concert with the basal ganglia. Damage to the basal ganglia is one of many causes that can lead to a loss of motivation (apathy): while the primary motor cortex still generates several programs, there is no arbiter to select the best one. Even with focus of the prefrontal cortex, this might not be sufficient to create a strong enough signal to execute an actual physical action.

> **EXECUTIVE FUNCTIONS** · The *executive functions* represent a series of means by which the prefrontal cortex can suppress thought patterns in other parts of the brain. While winners in the neural competition are selected with the help of the basal ganglia, the prefrontal cortex can counteract those decisions in favor of other actions. The prefrontal cortex makes these decisions based on its models. For example, stealing goes against the norms of society, so the prefrontal cortex suppresses the (utilization behavior of the parietal lobe's) urge to grab someone else's property.

As a compromise between focus, energy usage, and the possibility to quickly react to your environment, the brain actually *oscillates* between the parietal lobe (utilization behavior) and frontal lobe (focus) activity.[14] This way, while the prefrontal cortex focuses on the current action, the neurons in the parietal lobe are not firing and vice versa. A healthy prefrontal cortex would suppress distractions, or interrupt activities that go on for too long (in favor of being able to pursue multiple goals and react to the environment). Impairments to the prefrontal cortex and its communication with the rest of the brain can manifest as Attention Deficit Disorder (ADD or ADHD). It leads both to a lack of focus and to jumping between different tasks, as well as to "hyperfocus" when the afflicted person focuses on one activity for a long period of time.

In Chapter 5.5.1, we have learned that with a damaged prefrontal cortex, we would tend to show utilization behavior, which means applying action to anything in our environment that catches our attention (e.g., eating the candy bar). What initiates utilization behavior is the parietal lobe, based on what has been provided by the senses and processed in the occipital and temporal lobes. This is how most animals (and human children) react toward their environments if their prefrontal cortex is not as developed as it is in (adult) humans. When an animal is presented food, the animal eats it. When the ani-

---

[14]Misselhorn, Friese, and Engel, 2019.

mal is shown a toy, it plays with it. When we present a smartphone to someone with a short attention span, he might feel compelled to pick it up and check his messages, even if doing so is inappropriate in a given setting.

One can also imagine the executive function like a separate group in our neural committees. This new group takes care that, once started, activities are not stopped but brought to a finish. For example, eating the candy bar might be the most interesting thing, but finishing a work assignment might save us some time: interrupting work is often connected with a lot of overhead because you have to start thinking about it again from the beginning. Instead of going back and forth, our brain focuses on one thing, finishes it, then moves on to the next.

After an action has been executed, the thought pattern originally causing the action will be suppressed. For example, when standing in front of a sink and seeing a piece of soap, we might be inclined to wash our hands (utilization behavior). If our prefrontal cortex does not intervene, we will wash our hands. Once finished, our prefrontal cortex will then send a signal to suppress washing our hands a second time—it has evaluated our hands as now being "clean" and "safe." If this signal fails (which it sometimes does, for example in sufferers of OCD, see Chapter 5.5.1), the brain's internal status of the hands still shows that they are dirty, so we wash them again. Sufferers of phantom limb pain face a similar problem. Without the signal that the limb has been unclenched, the internal status remains unchanged. For both sufferers of OCD and phantom limb pain, one could argue that they could just look at their hands or missing limb. As we have discovered in Chapter 5.4.3 and Chapter 5.5.2, the brain requires two sources of information to update its body schema. In the case of phantom limb pain, it is the visual signal (with the help of a mirror) and the initiation of muscle movement (unclenching). In the case of OCD, it is the reevaluation of the risk together with the act of handwashing.

We can test this suppressive effect of the executive function by measuring the delay in reaction time between congruent and incongruent stimuli. For example, the word "blue" printed in red font has a different meaning than its color (incongruent stimulus). It is more difficult to process than the word "blue" printed in blue font (congruent stimulus). To experience it yourself, take a look at https://www.philosophy-for-heroes.com/blog/color-test and try to name the color of each word instead of reading the word. To give the correct answer, the prefrontal cortex has to suppress the part of the brain that provides the meaning of a word ("blue") in favor of the part of the brain that provides the information about the color of the text (which may be, for example, red).

6.3.5 The Inner Voice and Mind's Eye

Depending on the strength of the connections in the brain, the ability to imagine things can differ from person to person. For example, people with *aphantasia* do not have a "mind's eye." They lack visual imagination. They might still have *spatial* imagination to figure out where in the house to put a flower vase, but they cannot consciously imagine whether it would fit in terms of colors or form. They might have an abstract knowledge about what types of plants, colors, and vases they have in their home, but they cannot conjure a combined image and describe it.

This is similar to how we cannot imagine smells because the sense of smell is not connected to the system in the brain that creates imaginations. This is an indication that the sense of smell is wired differently when it comes to this inner experience. This is supported by the fact that the sense of smell remains active even when a person is otherwise unconscious in a coma because of a damaged thalamus.[15] We can also make memories of smells without consciously being

---

[15] Arzi et al., 2020.

aware of them. Both facts point to the olfactory system being able to bypass the thalamus and send signals directly into the neocortex.

For people with *hyperphantasia*, even textual descriptions can evoke a rich inner visual imagery, sometimes referred to as "mental cinema." fMRI studies suggest that hyperphantasia is related to a high activity in the visual cortex and low activity in the prefrontal cortex, while aphantasia is related to a high activity in the prefrontal cortex and a low activity in the visual cortex.[16] Similarly, it seems that a strong mental focus on the mind's eye or inner voice can suppress rTPJ activity and induce inattentional blindness (similar to hemispatial neglect or blindsight but not caused by damage to the brain) or deafness.[17] This is in line with our own experience where daydreaming can lead us to no longer pay attention, despite our eyes being open. This might be the reason people tend to look into the sky when thinking about something and using their imagination. This frees up processing power in the visual cortex as signals no longer have to compete with direct sense data.

People with hyperphantasia are aware that what they are experiencing are imaginations and not hallucinations. This is because their brain has the information that a particular imagination was initiated by themselves. This is different from people suffering from a tumour near their thalamus. It can generate nerve impulses that cause the experience of, for example, hearing voices. Without the information that the words were either created in Wernicke's area or spoken by another person, the brain can end up assuming that the voices must come from a supernatural source.[18]

In summary, it seems that the thalamus, visual cortex, and prefrontal cortex are involved in imagination. So, the surprisingly simple solution to explain the inner voice or mind's eye is that it is a descrip-

---

[16] Fulford et al., 2018.

[17] Todd, Fougnie, and Marois, 2005.

[18] Dutschke et al., 2017.

tion of anything transferred to the thalamus. Given that we are not constantly flooded with pictures or sounds (some people are—for example, the filtering mechanism for people on the autism spectrum is limited), this forwarding is regulated. A thought "pops up in our mind" if it hits a certain threshold—and this is arbitrated by the basal ganglia. In that regard, the "stream of consciousness" is a list of the most dominant thoughts at a given time. While this still does not answer *how* the thalamus can experience anything, we can update our diagram of consciousness (Figure 6.18):

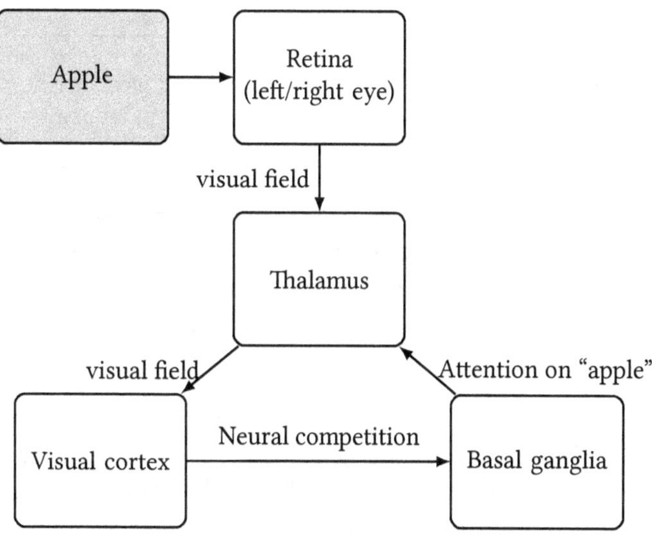

Figure 6.18: The process of perception from the object to the pre-processed image in the cortex, and back to the thalamus.

Logically, if we somehow shared our thalamus with another person, they could listen in to what we call our inner voice. We know at least of one case of conjoined twins (Krista and Tatiana Hogen) who share the thalamus but have separate neocortexes (see Figure 6.19).

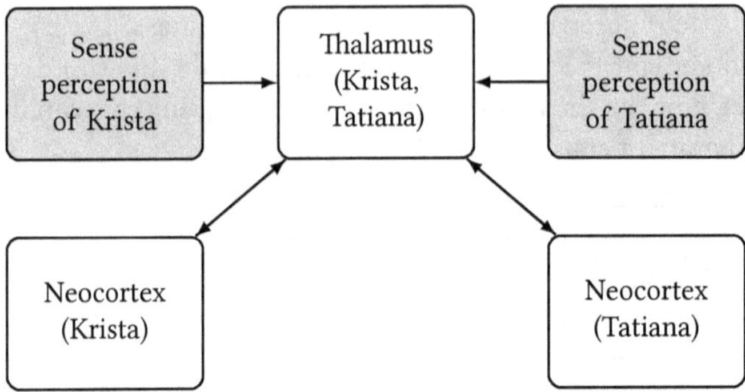

Figure 6.19: The conjoined twins Krista and Tatiana Hogen share the same thalamus, letting them share their perceptions and even thoughts with each other.

Besides being able to perceive what the other twin perceives with her sense of sight, auditory sense, or sense of touch, each twin can perceive what the other twin is thinking.[19] For example, what one twin sees is still processed in her own visual cortex. But as the sense data passes through the thalamus, it is also processed in the visual cortex of the other twin. Just like people with a damaged corpus callosum, the twins do not share a direct connection between their *neocortices*. In that regard, they only perceive the *results* of the thought processes of their twin—just as if they said out loud whatever came into their minds (thalamus). So, just because their loop of consciousness is connected, this does not mean that they actually think together on one problem like our two hemispheres do with the help of the corpus callosum. In a way, both hemispheres are like two singers. As long as they hear each other, they can easily

---

[19]Squair and Squair, 2012.

synchronize with each other. If they are disconnected, they lose the beat and can only coordinate with each other what song they will sing next. What happens during a "song" (loop of consciousness) is not synchronized.

Now, from the thalamus, the information is wired back into the cortex depending on the type of information (for example, imagining an apple activates the visual cortex). There, it again can take part in the neural competition and might again emerge as a winner. This forms the so-called *cortico-basal ganglia-thalamo-cortical loop* or, as we will refer to it in short, the *loop of consciousness*.

> **LOOP OF CONSCIOUSNESS** · The *loop of consciousness* refers to a network of connections in the brain from the thalamus to the cortex, then to the basal ganglia, and then back to the thalamus (also called the *cortico-basal ganglia-thalamo-cortical loop*). It is believed that this is the source of our subjective experience of the world.

This loop-like structure also explains why our inner experience of consciousness is *linear*: before we can think about a thought we just had, it has to be routed through the basal ganglia and thalamus to arrive back in the cortex. This also applies to split-brain patients whose left and right hemispheres are not connected by the corpus callosum. While the left and right side of the thalamus are conjoined, there are no direct nerve connections between the left and right hemispheres of the neocortex. In that regard, the split-brain patient's consciousness*es* are linear, too, but they have two separate loops. Someone with a split-brain syndrome is very similar to the conjoined twins Krista and Tatiana Hogen: there is no synchronization between the hemispheres, but the loop of consciousness of both hemispheres pass through the same thalamus (see Figure 6.20). The main difference is that each hemisphere of a split brain is incomplete, while in the case of the conjoined twins, they each have a left and right hemisphere.

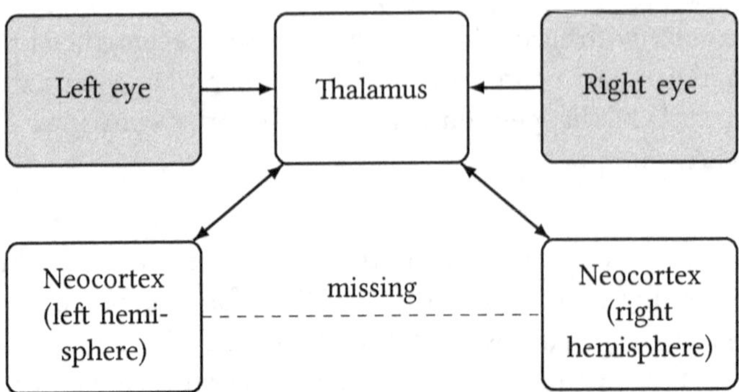

Figure 6.20: In a split-brain patient, the connections are similar to those of conjoined twins with a shared thalamus.

Having now gained an understanding of the role of the two hemispheres, the corpus callosum, the thalamus in creating the "inner voice" and "mind's eye," and how the working memory is filled, the next step is to examine more closely how the prefrontal cortex processes the information in the working memory.

6.4 Model-Building

We have learned how the brain routes (working memory) information from the neocortex through the thalamus and back to the neocortex to create the "inner voice" or "mind's eye." We have also seen how the working memory is used to transfer high-level information (like "the red apple is on the table") to the prefrontal cortex. Yet the still unanswered question is how consciousness would arise from that loop. Simply claiming that information passing through the thalamus results in our subjective experience is not a real explanation. To approach this issue, let us first examine when we think someone else is conscious.

Figure 6.21: The robot "Pepper" turns its head towards faces, creating the illusion of consciousness (image source: Shutterstock).

It actually requires very little proof for us to initially think of someone or something as conscious. Just having similar capabilities of the superior colliculus (tracking a moving object with its head and eyes) can be enough for us to automatically attribute a type of "consciousness" to a person, animal, or thing. An example is the robot "Pepper" (see Figure 6.21) which (who?) is programmed to recognize faces and move its (her?) head.

Now, to be *convinced* that the person (or machine) in front of us is conscious we need more than just observing the person's (or machine's) ability to follow an object with his (or its) eyes. In this section, we will look at the how a brain (or machine) can learn and adapt to the environment instead of just reacting in very predictable ways (for example, following an object with one's eyes).

6.4.1 Supervised Learning

What components are required for a machine that can learn and react to its environment?

For our examination of learning, we will look at a basic thermostat that regulates a room's temperature. It starts heating by switching on the heating unit when the detected temperature drops below a certain limit, and switches off the heating unit once the temperature reaches the limit (see Figure 6.22). Obviously, if the room temperature drops quickly or if the connected heating unit takes a long time to warm up or cool down, this might lead to significant deviations from the desired temperature.

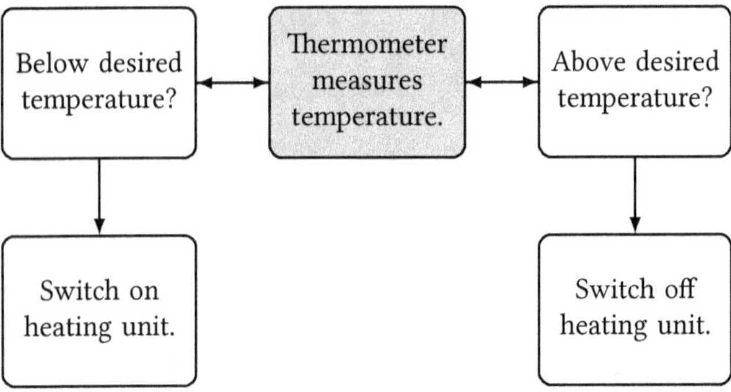

Figure 6.22: The programming of a basic thermostat that switches on if the temperature is below the desired level of warmth, and switches off once that temperature has been reached.

Figure 6.23 shows an example in which the thermostat is initially switched on (t_2), and heats the room and switches off when the desired temperature V_2 is reached. As the heating unit itself still emits heat after the heater is turned off, the thermostat will overshoot the target temperature. Then, it undershoots the target temperature because the heating unit needs time to fully heat up. It keeps over- and undershooting the target until the desired temperature (within the margin of error in the error band) is reached. If our brain worked like that, we would, for example, have trouble grabbing a glass of water as we would reach too far or not far enough.

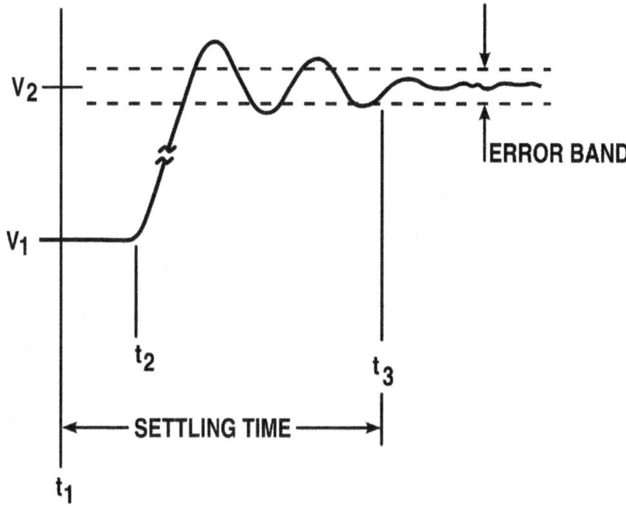

Figure 6.23: If a controller reacts only to external data, it tends to over- and undershoot the desired value (for example, a target temperature in the case of a thermostat).

To improve the programming of the thermostat, we can add a simple feedback loop which adapts internal values when noticing that the expected measurements are too high or too low. For example, the thermostat could use two variables to switch on its heating unit earlier (when it predicts the temperature to drop) or switch off its heating unit earlier (when it predicts it will overshoot the desired temperature):

- **Cooldown time**: This variable records the time it takes for the heating unit to cool down. If the measured room temperature ended up being too high after the heater was switched off, then the cooldown time was too short; the thermostat was switched off too late. The thermostat could increase the cooldown time a bit and perform better next time.

- **Heat-up time**: This variable records the time it takes for the heating unit to heat up. If the temperature ended up being too low, the heat-up time was too short; the heater should have started heating earlier or switched off the heating unit later. Again, the thermostat could adapt the heat-up time.

With the cooldown timer and heat-up timer in place, we have created a basic system that can learn over time (see Figure 6.24). The thermostat could keep the room temperature closer to the desired value (and save energy) by starting or stopping heating earlier or later. For example, if the desired temperature is 65°F, the thermostat probably has to switch off the heater some time before the room temperature reaches 65°F because the heating unit is still hot when switched off.

Using variables and adapting them depending on the outcome of an action can be applied to problems of any size. As it requires interaction with and feedback from the environment, this type of learning is also called *supervised learning* as some sort of "supervisor" is involved to decide whether or not a decision was productive. In the case of the thermostat, the supervisor was the function that checked whether or not the desired temperature has been reached ("Did the temperature in the room rise above the desired value?" and "Did the temperature in the room fall below the desired value?", respectively). If the desired temperature was missed, the thermostat failed in its process and the heat-up or cooldown timers need to be adjusted.

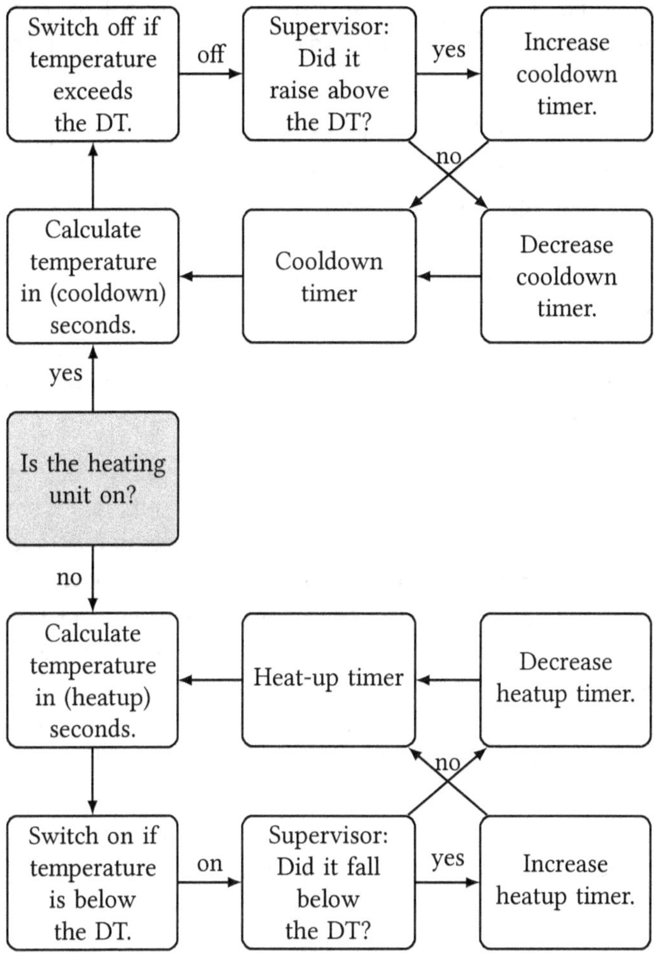

Figure 6.24: A thermostat that aims to raise the room temperature to a desired temperature (DT). It has a feedback loop that allows the adaptation of a cooldown and heat-up timer after switching off the heater reduces the settling time.

SUPERVISED LEARNING · Using *supervised learning*, a brain (or computer) can improve its response to a situation with each new encounter. For example, a dog can learn to sit or roll over on command by getting positive rewards for doing so during training.

The same principle that applies to the thermostat also applies to, for example, learning to throw a ball at a target. The brain records which chain of neurons are responsible for an action. If you observe yourself missing the target, the erroneous chain of neurons is weakened and you might throw the ball differently the next time. If you hit your target, the successful chain is strengthened and you will be more like to throw the ball the same way the next time. That means that the next time, you are less likely to miss or more likely to hit the target, respectively.

The major downside of supervised learning is that it always requires interacting with the environment and observing the outcome. The two variables represent the behavior of the thermostat (when to start or stop heating), not the behavior of the room's air temperature. Hence, while the feedback loop of the learning thermostat we discussed above reduces the time to reach a stable temperature in a room, it requires a trial-and-error approach. As such, a system based on trial and error is like a black box into which you cannot look. You could neither open it to learn the reasoning behind each activation or deactivation of the heating unit, nor could you tell the thermostat to update its heat-up and cooldown time according to the new coordinates, room size, or window configuration.

In our brain, one of the "supervisors" is the amygdala, which has mapped sense data to positive and negative emotions, with positive emotions strengthening a chain of neurons, and negative emotions weakening it. Vice versa, the amygdala itself learns this mapping of sense data to positive and negative emotions with its own supervisor. The amygdala's supervisor is the experience of physical pain or pleasure, a hardwired mechanism in our body. In turn, pain and pleasure evolved as each proved to be advantageous for the fitness of lifeforms to prevent damage and encourage procreation and locating food. In summary, decisions of our neocortex are supervised by the emotional mappings in our amygdala. The amygdala learns those emotional mappings with the help of the pain and pleasure mechanism of the body, which in turn evolved over generations by selection.

The neurons in the amygdala (and in other parts of the brain) learn by what is called "backpropagation," which is comparable to a bucket brigade where items are transported by passing them from one (stationary) person to the next. This method was used to transport water before hand-pumped fire engines; today, it can be seen at disaster recovery sites where machines are not available or not usable. To encourage this behavior, everyone in the chain is later honored, not just the last person of the bucket brigade who is actually dousing the fire. In neuronal learning, the last neuron back propagates the reward from end to start, allowing for the whole "bucket brigade" or chain of neurons to be strengthened. This encourages similar behavior in the future—it is the core of learning.

A good example for supervised learning is training a dog to sit on command. The learning cycle is as follows: First, you give the dog treats. Using the pain and pleasure mechanism, the dog's amygdala connects the sense data (receiving a treat) with a positive emotion. She has learned that your treats are good. Next, you let her stand while you wait until she sits down. While waiting, you repeatedly tell her to sit. Of course, she does not understand what you are saying. Eventually, she will get tired and sit down. Then, you give her a treat. Her amygdala will link the command "sit," sitting down, and receiving a treat. In time, she will feel happy about not only receiving a treat, but also sitting on command.

What would happen if we programmed such a system into a computer and gave it a language module to speak with us?

6.4.2 Turing Test

In the wake of the development of computer technology, Alan Turing asked this question in 1950:[20] How could we test whether a computer is as intelligent as or indistinguishable from a human? He proposed the so-called *Turing test*, in which a person is put in a room with a computer terminal. Using only a textual chat interface, the person has to figure out whether or not his chat partner is a computer program or a human being. If the person interacting with the computer cannot give a clear answer (or is convinced that the computer program is human), the machine has passed the Turing test.

> **TURING TEST** · In 1950, Alan Turing proposed the *Turing test* to assess whether or not a machine is intelligent. In the test, a human participant would observe a text chat between a computer and a human. The machine would pass if the observer could not tell who was the machine and who was the human.

What questions would you ask to determine whether you are talking with another person or a computer?

You could run through questions of a basic intelligence test but that would not answer the question of whether the computer has human-level intelligence and a sense of "self," only that it is intelligent. Likewise, if you straightforwardly asked whether or not your chat partner has an inner experience, the computer could simply respond with a programmed response of "Yes, I am conscious of my inner experience." With this line of questioning, it seems that all we can find out is how well-prepared or well-programmed the other side is, not whether or not the chat partner is conscious.

It seems that our intuitive understanding of consciousness depends primarily on the observer attributing consciousness to the person or

---

[20] Oppy and Dowe, 2019.

entity. Given that with the Turing test, the arbiter of whether or not a computer is conscious is always the subjective opinion of a test person, this is not surprising.

Because answers to pre-determined questions could be prepared beforehand, we could test the flexibility of our conversation partner by referring to the conversation itself. For example, we could talk about food preferences we like and then ask the chat partner what dish it (he?) would recommend. To answer this question, there is no general rule as each person is different. The machine (or person) would have to evaluate our preferences throughout the conversation to build an idea about what we like or dislike. This is similar to programming a chess computer: you can program the opening moves into the computer, but once the game diverts from the database of memorized positions, the computer needs to rely on actually playing chess and predicting your moves. For example, we could bring up a statement like "When you ordered me a pepperoni pizza yesterday, I told you how great it was. I lied. I actually prefer pizza with tuna."

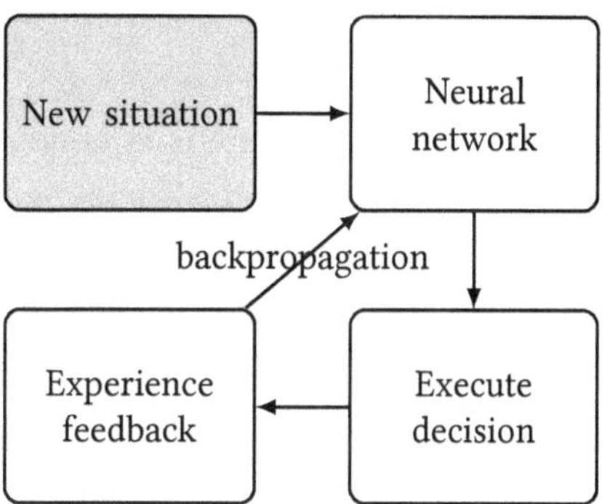

Figure 6.25: With supervised learning and backpropagation, the brain can react to new information only through supervised learning after making a decision and having experienced the consequences.

A computer program with supervised learning based on neural committees with backpropagation (see Figure 6.25) could not do anything with updated information during a conversation. It could only correlate two different sense data and evaluate whether or not that response was positive or negative. For example, it could take into consideration your immediate positive reaction when it had ordered you a pepperoni pizza. However, modifying its own neural committees based on specific new information (that you like tuna instead of pepperoni) is impossible as it is a black box. All it could do is to evaluate its action of ordering you a pepperoni pizza *in general* as negative and then (hopefully) find out through trial and error that you actually like tuna. It could not order you specifically a tuna pizza because it can only learn through action and reaction, not through abstract information. It could only decide to order you another random pizza. This of course sounds anything but intelligent. Why could the computer not just order a tuna pizza? Well, the same question could be asked about the thermostat: why could the thermostat not just understand that we moved it to another room?

The challenge is that we are quick to attribute abilities to a machine that it does not have. Just because the computer could speak to us does not mean it can react to information like we do.

In summary, supervised learning alone is not enough to react flexibly within a conversation and pass the Turing test. This leads us to a different approach to learning, which does not need direct interaction with the environment, namely the so-called "unsupervised learning" method.

6.4.3 Unsupervised Learning

How can we gain new knowledge about the world without relying on trial and error?

The difference between supervised learning and unsupervised learning is that the former requires trial and error, while the latter just needs sense data to build a model of the world. Scientists developing automated machines have shown that to control a certain variable, *"every good regulator of a system [needs to run] a model of that system."*[21] For example, to keep the room at a certain temperature (the variable), the thermostat (the regulator) needs an internal model of the system (the heating unit and the room it heats—the system).

Our initial action depends on whether or not we are familiar with the involved entities. When encountering a new object, we are first inclined to go around it, touch it, or put things on it until we have formed a basic idea what that object is. As discussed in *Philosophy for Heroes: Knowledge*, we create the concept of, for example, a table by looking at many different tables, ending up with properties like number of legs, size, material, and shape. When deciding whether or not a table would fit into our living room, our brain relies on the spatial understanding of the particular table. For this, our brain adds the specific measurements or values of the properties of the concept, forming a *model* of the table in our mind.

> **MODEL** · The *model* of an entity is a simplified simulation of that entity. It consists of the entity's concepts and its properties, as well as some of the entity's measurements.

---

[21]Conant and Ashby, 1970.

We have already discussed examples for the brain building such models, for example, of the external world (object permanence), a body schema, as well as the inner world of yourself and others (theory of mind). The body schema answers the simple question "Where are my limbs?" Without a body schema, we could still do anything we can do without it, but it would be much harder. This becomes apparent when we, for example, try to manipulate something with our hands when our vision is warped. When we can observe our hands only through a mirror, we have to translate our movements consciously (left becomes right); we cannot rely on our body schema and simply grab an object.

In principle, mammals use some sort of body schema to incorporate changes in their physiology (body size, limb length, etc.) as they are growing up. Similarly, we have learned in Chapter 5.4.1 how the superior parietal lobe maintains a mental representation of the internal state of the body. While the current state is provided by the internal and external senses, it is significantly more accurate to have an internal model that is constantly updated by the input from the senses.[22]

We see the same model-building when using tools. As we have learned in Chapter 5.4.1, our brain sees tools as temporary extensions of our limbs. To do this, our brain manages several different body schemas (models), one for each type of tool. While humans are typically able to do this, many other animals have difficulties with this task. For example, dogs often fail to incorporate things into their body schema. Let us consider a dog with a stick: when holding a stick in her mouth, she might struggle to get through openings because, in her mind, the opening is wide enough for her body, but she might fail to include in her calculation the space needed for the stick. There are two possible ways for the dog to deal with this situation. One way would be that she learns through trial and error to tilt her head when walking through an opening while holding a

---

[22]Wolpert, Goodbody, and Masud Husain, 1998.

stick. The downside of this approach is that it might work for some openings and sticks but not for others. Whenever she faces difficulties, she would again have to rely on trial and error until she turns her head in a way so that she can pass through with the stick. A better way would be that she creates a model in her mind of herself with the stick in her three-dimensional environment. This way, she could understand why turning her head is important, and to what degree she has to turn her head depending on the dimensions of the stick and the opening through which she is moving.

UNSUPERVISED LEARNING · Using *unsupervised learning*, a brain (or computer) can build a concept by analyzing several sense perceptions, finding commonalities, and dropping measurements. For example, unsupervised learning could be used to form the concept "table" by encountering several different tables and finding out that they share properties like having a table-top, the form, material, and size of the table-top, and the number of table legs.

Our brain uses both supervised and unsupervised learning, one to decide upon an action, the other to model the world and make predictions. We use unsupervised learning to create a model of the world to predict what will happen, evaluate it, and feed it back to our supervised learning as "reward" or "punishment."

Applied to the example of the thermostat, we would know how quickly it cools down, how long it takes to heat up the whole room, differences between summer and winter, habits of the people living there, etc. during installation. With this information, the thermostat can calculate the optimal time to switch the heating unit on or off (see Figure 6.26) without having to learn it by trial-and-error, significantly speeding up the progress of adapting to new environments.

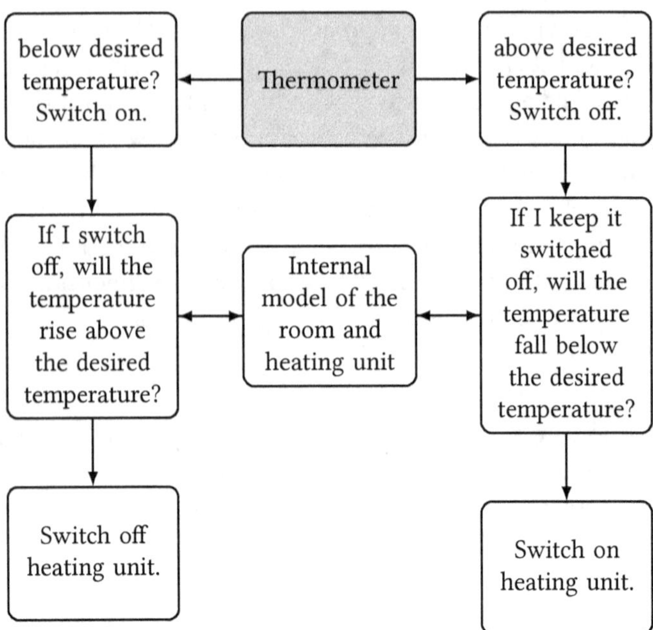

Figure 6.26: Intelligent thermostat that tries to predict the future temperature with a model in order not to overheat the room or start heating too late.

But the approach of having a model of a situation to which you can apply parameters allows more than just quicker learning. It enables you to provide *reasons* for your decision and explain it to others using your model. For example, in the case of the thermostat, it might suddenly start the heating unit during the day even though the sun is shining. Without additional information from the thermostat, we might start wondering if it is defective. In such a case, a smart thermostat could inform us, for example, that the window is open or that the weather service reports an upcoming snow storm. We could also tell the thermostat that we have just installed automated blinds that open in the morning (allowing the sun to heat up the room). The thermostat could take that information, update its model and start heating the room correctly (maybe with some minor adjustments) on the first day without having to spend weeks of learning the new environment.

The thermostat could even start giving us hints about how to reduce the energy bill. This requires an *understanding* of the underlying model, and the ability to translate it into language. Basically, this refers to the ability to teach the user a simplified model the thermostat itself is using. Ultimately, a thermostat can only be seen as being as smart as we are if it is able to program ("teach") another thermostat. In a social setting, this would be the equivalent of explaining our own behavior to others. If you are, for example, at a lecture and all your brain is doing is to make you jump up and rush home, what would the people around you think? Instead, if you are conscious of the steps leading up to the decision (for example, remembering the electric iron you forgot to switch off, combined with a self-image of being forgetful), you can provide a reasonable explanation for your behavior.

Another example for an application of unsupervised learning is in classifying images using artificial intelligence. A computer system based on unsupervised learning still requires that it is presented with inputs, but it does not have to interact with its environment to adapt to new (but similar) conditions. It simply learns how to *describe or differentiate* the input according to a number of variables, not to make actual decisions. For example, if we provided the system with pictures of faces, with the unsupervised learning method, it would return variables relating to age, sex, hair color, facial expression, eye color, facial geometry, and so on. This approach can be powerful because when encountering new objects that it has not yet observed (but that fit into its schema of properties), it can derive other properties from them. For example, if it has learned how faces change with age, it can make predictions about how the face might have looked in the past or will look in the future.

In summary, it could be said that unsupervised learning *allows you to imagine how things would be in a different situation.* In the case of the thermostat, we could "ask" it about its heating plan if we were to move to the North Pole, the equator, the basement, or to a room with a lot of windows. With some additional programming, we could even inquire, for example, the heating costs we would save by moving to a smaller house or apartment.

Applied again to humans, it seems clear that those of our ancestors who had more in-depth insight into the inner workings of their minds were able to form better relationships. Being able to arrive at logical decisions and to communicate how they came to those decisions made them reliable members of society. Others could better predict how they might act—assuming the explanations are not created after the fact as rationalizations for their behavior.

6.4.4 Summary

Let us now apply what we have learned to the Turing test and the example of ordering a pizza. If the machine had an internal model of your preferences, it could recall the situation when it originally ordered you the pizza, include your new assessment of your preferences, and run another backpropagation on the original neurons that led to the decision. This way, it would lead to a higher evaluation of your preference for tuna pizza, and a lower evaluation of your preference for pepperoni pizza. And *that* is the power of learning using a model (see Figure 6.27) versus trial and error. Once we have conceptualized a situation and learned how to handle it, we can easily apply the knowledge to similar situations with different parameters (measurements). And in the context of a conversation, we can directly incorporate new information in our response.

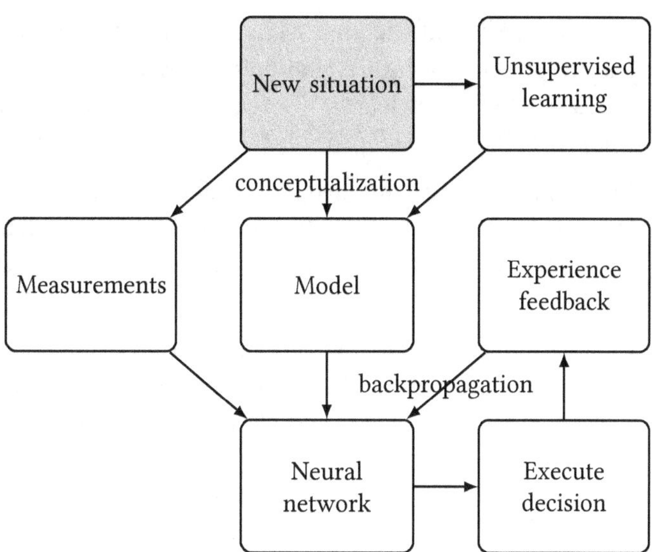

Figure 6.27: With a model, a neural network like the brain can learn (adapt its model) without actually having to test the decision against reality.

This concludes our examination of the elements of consciousness. It is now time to collect our findings and form a coherent theory of consciousness. So far, we have learned:

- In Chapter 5.1, we learned how different brain parts evolved over time, and how they relate to decision-making. We still need to take a look at how each step contributes to evolutionary fitness. Our goal is to determine what use consciousness has for us.

- Chapter 5.2 led us through our evolution, comparing humans to other primate species. With what we have learned in the other parts of the book, we still need to precisely point out the differences between humans and apes when it comes to consciousness (assuming there are any).

- In Chapter 5.3 and Chapter 5.4, we discussed how the brain tries to predict the future, and how it builds a body schema. Similarly, the brain learns to create a theory of mind (see Chapter 5.5).

- In Chapter 6.1, we found that it makes the most sense to look at consciousness from a monist perspective (materialism). This helped us to draw a diagram explaining the process of consciousness step by step.

- Chapter 6.2 covered what effects various cognitive defects can have on our perception of self. We clarified the difference between a lack of awareness and blindness and discovered that it is not only our senses that are pre-processed, but also our attention. We cannot react to things we are not being made aware of by our brain, but we can learn strategies to overcome such a lack of consciousness.

- In Chapter 6.3, we learned what the "inner voice" and "mind's eye" are, and how these relate to the working memory and the prefrontal cortex. This also allowed us to clarify our understanding of the process that creates consciousness. The remaining question was where the subjective experience ultimately comes from.

- Finally, in Chapter 6.4, we saw how the brain can build models of the world and in what way they are advantageous to us. We tied this in to our discussion of concepts in *Philosophy for Heroes: Knowledge*.

In summary, what is still missing is an explanation for the subjective experience of consciousness and a discussion of the evolution of consciousness in comparison to that of apes. In the course of this discussion, we will develop a new theory, the *awareness schema*, which we will discuss in the next section.

6.5 The Attention and Awareness Schema Theories

The *attention schema theory* was developed in 2012 by Michael Graziano, a neurobiologist from Princeton University.[23] It states that beyond the models of the body (the body schema), the mind (theory of mind), and the world (object permanence), the brain has also evolved to build a model of how the brain *itself* selects and focuses on certain sensory information. The internal model of this process of attention is called the *attention schema* (the *model* of attention). Just like we can use our body schema intuitively (we decide to grab a glass of water instead of thinking about how each individual muscle has to move), we can use the attention schema intuitively to access what we are currently thinking about. In our discussion, we have already encountered the parts making up the attention schema. We referred to them as the working memory, theory of mind, and the stream of consciousness.

> **ATTENTION SCHEMA** · The *attention schema* is a model the brain creates of the process of attention. It allows access to the working memory in order to be able to intervene before an action is taken.

When we are being asked what we are thinking, we cannot reply by listing the biochemical state of every neuron. We can access our model of attention (the working memory), though: what is the current subject or thought on which our brain is focused? This is a brief summary of the results of the selection processes going on in our brain. For example, when we are looking at an apple, our visual system will recognize the concept of "apple." By accessing the properties connected with the apple (an apple is round, has a certain color, has a stem, tastes sour to sweet, and so on) and the information from our visual system about the apple's measurements (color, size, location, etc.), our brain can create a *model* of the apple.

---

[23]Graziano and W Webb, 2015.

If we are asked to grab the apple and explain our physical relationship to the apple, we access our body schema and the model of the apple. Our working memory contains something along the lines of "I am grasping this red apple with my hand, while my arm is outstretched." When we are asked what we are currently conscious of, our brain interprets the contents of the working memory as "I am paying attention to the apple." The "I am paying attention to" part is an interpretation of the brain's process of selecting thought patterns (the neural competition), the "apple" part is an interpretation of the most prominent thought pattern. To summarize, to answer the simple question "What are you conscious of?" we need to access two models:

- The model of the apple ("What properties does the apple have? Where is the apple in relation to me?"); and

- The model of attention ("What am 'I' paying attention to and what is the meaning of 'I' in this context?").

Further self-reflection reveals that we do not have any direct access to any details of how this attention works. We cannot describe intuitively how we decided to look at the apple. Just like we do not intuitively know how we can produce speech with our muscles and tissues, we have no idea about our brain architecture just by thinking about it. Our answer might simply be "Well, *I* have decided to look at the apple, so *I* did, and *I* am perceiving the apple." But what exactly is the "I"? Depending on our upbringing, we might say that this is our soul, mind, or consciousness, without being able to go into further detail.

And just like we cannot access how this awareness of the self works in other people, we are puzzled by this phenomena in ourselves, too. This intuitive understanding of attention stays with us because our model of attention is built the same way we build models of, for example, the spatial dimensions of objects, our own body schema, or models of our own or other people's minds (theory of mind). It is not a belief we can explain like a religious belief or scientific theory. Instead, it is on the perceptual level similar to optical illusions. We can know on an abstract level how our brain works, but that has no effect on our feeling of "self." The way we experience consciousness will stay with us, no matter how much we learn about it—just like we will interpret a combination of red, green, and blue light as "white light," no matter how much we learn about the electromagnetic spectrum.

The model of attention is not the same as attention, just like the model of a candybar is not sweet. But we are conscious only of that model. It describes something that is actually happening (the process of attention). The actual mechanics of what is happening differ from the simplified model we build intuitively, just like the actual mechanics of our limbs are much more complex than our simplified body schema. We can learn about all the muscles in the arm, but when raising our hand, we will intuitively raise our arm and hold up our hand, not think about activating a dozen muscles that make an arm and hand rise. Or when we want to sing higher, we do not think about the muscles in our throat, we think about singing higher. We intuitively access our simplified body schema.

With the help of the model of attention, we are able to access what is currently going on in the neural competition of the brain. This can help us to make better decisions by enacting corrective measures before actually making a decision and having to face its (potentially negative) consequences. When you feel angry and want to punch someone, your prefrontal cortex can suppress the action and divert it into something more socially acceptable (and less potentially dangerous) like cursing. This is better than seeing yourself acting out those emotions and trying to fix the situation afterwards.

At a neuronal level, the attention schema is very much like the smart thermostat we have discussed in Chapter 6.4.3. You can compare the temperature curve of a room with the amount of cognitive resources the brain uses to process an external signal over time. You do not want the room to heat up beyond the desired temperature, just like you do not want to have external signals suppress other brain activity. For example, if you are driving at night, a sudden noise might startle you, but you nonetheless want to keep some resources of your brain reserved for operating your car. You do not want to immediately forget the noise either, but maybe think about it and then take action. For example, maybe you recognize the noise as a burst tire and now decide to find a place to stop the car.

The thermostat-like ability to control the attention to a signal was confirmed in a study comparing the reaction time to visual cues we are aware of with those we are not aware of.[24] In the experiment (a so-called "Posner cueing task"), participants look at a computer screen and fixate on the center of the screen. On the screen, there is a box on the left and another box on the right side, with shapes showing up in them. The participant has to press a button once he or she is aware of the shape and the reaction time is measured. By displaying a cue (a dot or small arrow) for a very short time before the shape is shown, the reaction time can be reduced, even though the person did not consciously notice the cue. The study ultimately showed that attention and awareness are two different systems, and that we can better control signals that we are aware of (see Figure 6.28). This in turn means that awareness has a specific function and is not just a byproduct of evolution.

---

[24] Webb, Kean, and Graziano, 2016.

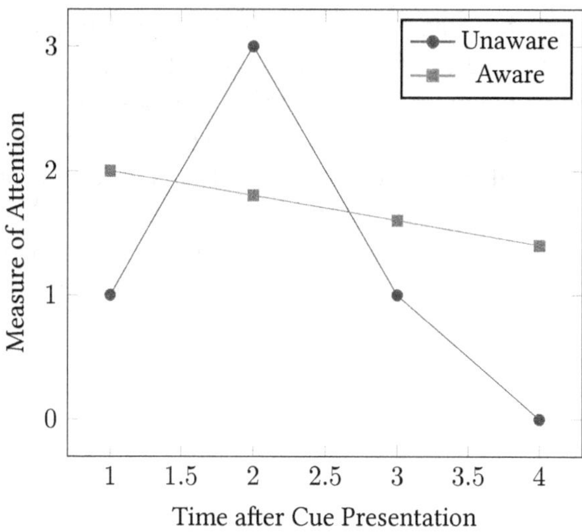

Figure 6.28: Comparison between the measure of attention after cue presentation of aware and unaware subjects.

6.5.1 Virtual Prefrontal Cortex

To understand how the attention schema works within the brain, let us imagine that we are playing a computer game in which we control a virtual character. Let us assume that this control is limited insofar as we can make the character do only things that the character's prefrontal cortex can make the character do. That means we are not steering that person from a first person or even third person perspective. Instead, we have only the ability to suppress or promote actions provided by the working memory by sending signals to the basal ganglia. For example, if we read the entry "I see a candy bar" in the working memory (see Figure 6.29), we can suppress the induced utilization behavior by the parietal lobe and resist eating it. For the decisions about which actions to suppress or to promote, we use our own brain. This allows us to at least to do anything that person is able to do based on what catches his attention. For example, we cannot make the person call a specific number that only we know, but he might see something that reminds him of his mother, leading to a possible desire to call her (which we could promote or suppress). Similarly, we cannot make him speak in a foreign language if he has not learned it before. A passage in a book might catch his attention and we can decide whether or not he will say it out loud, but we (as the prefrontal cortex) cannot make him out of the blue say things of which he is not aware.

Number	Origin	Working memory
1	visual	candy bar
2	visual	telephone
3	memory	mother's telephone number
4	visual	writing "Käsebrot"

Figure 6.29: Examples of the contents of the working memory of the virtual character. The virtual prefrontal cortex we are controlling in the game can tell the character to interact with the candy bar or use the telephone to call his mother, but it cannot understand the German word or call a number only we (the player) know.

In this scenario where we control a virtual character through his prefrontal cortex, what effects would (simulated) neurological conditions like hemispatial neglect have on our interaction with the virtual world? We cannot act upon things that his TPJs did not properly put into the working memory. We might initially not even notice that the character we are controlling in the game has hemispatial neglect as all we are getting in terms of information is the working memory. If we, after a number of encounters with other people and situations, discover that he has hemispatial neglect, we could actively train him (the rest of his brain) to turn his head by rewarding him whenever he does, allowing us to access the option to turn his head more often. Similarly, we could reduce the effort needed to exercise by training him that going for a jog is just part of the normal daily routine. We could not simply tell him to "start to exercise."

Instead of deciding which element of the working memory to act on in each instance, we can automatize our play by letting the computer decide what to do with the help of models. For example, when we see an apple, we might not want the character to eat it. Maybe we want him to focus on another, more important, task or to realize that he is not really hungry.

We could also expand his prefrontal cortex with social models that contain our ethics to automatize interactions. This could range from abstract rules like "do as you would be done by" to specific rules depending on the situation and people involved. Whether or not a particular social behavior is then promoted or suppressed would depend on that model. Similarly, we will do this with models like risk assessment, long-term planning, and so on. Eventually, he will act within the virtual world as an autonomous avatar.

6.5.2 The Awareness Schema Theory

While a compelling idea, the approach of understanding the brain by programming an artificial prefrontal cortex works well only if the character makes decisions about a present situation. But once he has to take into account the consequences of his actions, he has to imagine the future. The challenge is that each individual prefrontal cortex module cannot do this on its own. For example, when planning to climb up a wall, you have to combine the brain's resources of the frontal, occipital, and parietal lobes to imagine yourself and your limb positioning for each step of your climb. For this, it is not enough to simply read the working memory and suppress individual entries. Instead, the prefrontal cortex has to initiate the imagination of a future scenario where you are higher up on the wall:

1. Visually identify the first possible step.
2. Check with your parietal lobe whether you can physically reach it.
3. Forward the information to the working memory as a possible next step.
4. Suppress the movement by the prefrontal cortex.
5. Check whether the movement would fit into your plan to reach the top.
6. Return to the working memory the idea of taking this step to reach the top.
7. Imagine a future scenario where you have already taken the step.
8. Visually identify the next possible step.
9. Return to 2.

In other words, the prefrontal cortex recruits other parts of the brain to help with calculation. Those other parts process the data no matter whether they originate from the senses or from a thought-up scenario created by the prefrontal cortex. When you imagine sitting in the grass in the summer and enjoying the view of the mountains, that idea will be just as real at least for some parts of the brain as actually sitting in the grass. This of course depends on your prefrontal cortex' ability to recruit the visual cortex to create such an image.

Playing chess is another example where you have to "think through" the consequences of your choices before you act. You analyze your current position on the board, have a vague intuition of where to attack, and then calculate move by move how this strategy would turn out for you. Similar to climbing the wall, you would suppress the utilization behavior of just executing the first move that comes to your mind and instead store it in your working memory ("White moves pawn to e4"). But just like not wanting to think only of the first step when climbing, you do not want just to move the chess pieces, you want to think through a particular sequence of moves, one dependent on the other. This means that the output of your analysis of the first move needs to be the input for thinking about the second move, and so on. To achieve this, you would loop through several steps and imagine possible outcomes of the move until you either dismiss it or find that it will improve your situation. Chess players are good at evaluating whether a certain position is generally good, so they can tell relatively easily if another position is better or worse. A simple rule of thumb would be counting the number of pieces you still have on the board versus the number your opponent has on the board. The more chessboard squares you control directly or indirectly with your pieces, the higher your chances are to mount a successful attack.

One difference between climbing a wall and playing a game of chess is that you also have to imagine your opponent's moves. This makes the game challenging as you also have to use your TPJ and put yourself into the shoes of your opponent and think of possible strategies she might use to win the game. This makes it even more difficult to suppress what your eyes are seeing as you have to mentally rotate the board or remind yourself that her pawns are moving in the opposite direction of yours.

Without the possibility of looping your thoughts through the working memory, playing chess would be much more difficult. Your cortex would have to arrive at a good move in only a single thought-step. While the evolutionary competition among the committees could still come up with a solution to the problem, you would be unable to divide the problem or interrupt thinking about it to return to it at a later time. Learning chess would then require a much more intuitive approach, similar to what you need with the board game "Go" that is far too complicated to calculate deeper than a few moves.

The theory I propose is that the brain cannot just build a model of our attention, but also of our *awareness*. While attention refers to the neural competition in the brain, awareness refers to the loop of consciousness. Having an awareness schema would allow the brain to modify the contents of the working memory in order to imagine things, recruit other parts of the brain, or think about future scenarios step-by-step. As opposed to the attention schema ("What was put into the working memory?"), the awareness schema addresses the question of how entries in the working memory will stay there.

Modifying the working memory enables the core functions of the prefrontal cortex, namely to focus and pursue goals. Someone without an awareness schema could still use the prefrontal cortex' abilities to suppress behavior that does not lead to the goal, but he could not pursue goals that are independent of his immediate environment. The utilization behavior of the parietal lobe does not go as far as making use of the models of the prefrontal cortex to think about possible actions in the future; it merely analyzes objects that are immediately available. For example, the function (from our parietal lobe's perspective) of a piece of chocolate is to be eaten, or the function (from a cat's parietal lobe's perspective) of a glass is to be pushed off the table.

> **AWARENESS SCHEMA** · The *awareness schema* is a model the brain creates of the process of awareness. With the awareness schema, the brain can write to the working memory to influence what the brain will focus on next. It also allows the brain to imagine alternative, past, or future scenarios.

Adding this idea to our diagram results in Figure 6.30. There we see that the prefrontal cortex uses the working memory as an input for its models. The prefrontal cortex returns the results of this calculation to the working memory which in turn feeds it back into the loop of consciousness as an input for the next calculation. This allows the brain to not just suppress external distractions (for example, birds chirping outside), but also to imagine objects in a different condition or scenario. For example, we could imagine an apple in a bowl on a table, or hanging from a tree, or cut into pieces.

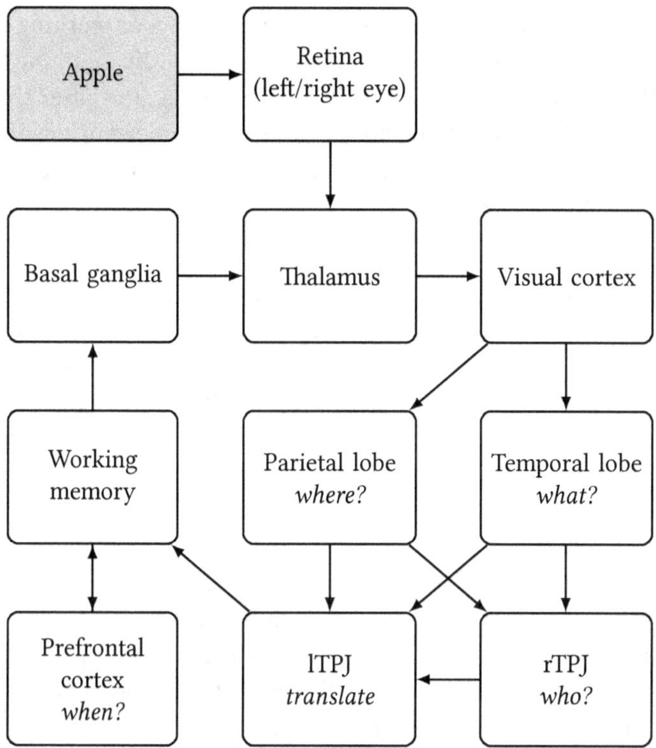

Figure 6.30: Adding input from the models of the prefrontal cortex to the working memory allows the brain to fill the working memory with predictions about the future or imagined scenarios.

Without the awareness schema, you could still pursue strategies by suppressing alternatives, but you could react to the consequences of your actions only after you have executed them. Similar to the thermostat with supervised learning (see Chapter 6.4.1), you may under- or overshoot your desired goal.

In that regard, this ability to modify contents of your working memory is a form of control. To accomplish this modification, the brain uses the awareness schema like the body schema. The latter enables you to intuitively use your body, while the former enables you to intuitively use the loop of consciousness and the mechanisms of focusing on particular thoughts.

Similar to how we can think about the past and future with the help of the awareness schema, we can manipulate the working memory in other ways. For example, when thinking about an alternative scenario (like we did in the example with the climbing), we could imagine not only the external world being different, but also ourselves being different. Try to imagine how people would react to you if you were very muscular or very fragile, and your prefrontal cortex would modify the model you have of yourself. You could even imagine yourself being another person, or you could imagine how other people would react in such a situation. For each case, the underlying principle is always the same: modifying the working memory to make the rest of the brain think about solving this alternative scenario. For example, if Anna thought about Peter joining her climb, she would run through similar steps of thought. But to imagine Peter climbing up the wall, she would need to replace the "Anna" entry with "Peter" in her working memory and think about what steps he would take.

With the awareness schema, we can image how the world might look from another person's (or animal's) view. This enables us to have empathy, and to predict other people's actions. However, damage to the system providing our awareness schema can seriously confuse our thinking process, resulting in, for example, us projecting our own emotions onto others. In contrast, lower brain functions *always* assume that a perception or emotion is from "self," but they lack the ability (at least without the neocortex) to imagine what others perceive or feel.

A possible theory of how imagination evolved is that it originated as an emotional coping mechanism discussed in Chapter 5.5.1. To imagine things, we need to manipulate our working memory which in turn activates, for example, our visual cortex. And the better we can manipulate our working memory, the better we can manipulate our amygdala because the amygdala cannot differentiate very well between "real" sense data and imagined sense data. If we can show emotional restraint and channel our emotions in the right direction, we are better at forming relationships and living peacefully within a tribe—a tribe that can aid us when we are hurt, help us raise our children, or protect us when we are in danger from predators.

6.5.3 Levels of Consciousness

Before we further refine our understanding of consciousness, let us summarize the different models the brain uses:

- **Object permanence**: a model of the world—"Where are which things in the world that do not belong to my body?".

- **A body schema**: a model of our body—"Where are which limbs?" "What belongs to my body and what is not part of my body?".

- **A theory of mind**: a model of other people's minds—"What do other people know?".

- **An attention schema**: a model of our own process of attention and that of others—"What am I currently focused on?".

- **An awareness schema of self**: a model of our own process of awareness and how we can manipulate our awareness—"What did I focus on, how can I change my focus, and what should I focus on next?".

We also create an attention schema of other people and animals. The difference compared to the model of attention we create of ourselves is that the sources of information are different. We infer the focus of another person by his gestures, facial expression, eyes, and what he is talking about. For ourselves, our brain can create a richer model of attention because it also has direct access to brain processes.

To improve our understanding of consciousness, we can rank our different abilities and schemas in levels:

1. **Sense data**: Sense data is registered and pre-processed in the brain.
2. **Attention**: The ability of the brain to focus on a particular set of sense data using neural competition.
3. **Awareness**: Something that was put into the working memory including information about "what," "where," and "who."
4. **Attention schema**: The ability to access the working memory to use it to update the prefrontal cortex' models and use those models to suppress or promote individual actions.
5. **Awareness schema**: The ability of the brain to manipulate awareness to think about alternative scenarios. The working memory is used to manipulate future loops of consciousness.
6. **Philosophy and science**: Abstract knowledge about awareness, enhancing the brain's ability to create habits and strategies.

Information flows mostly in one direction from level 1 (sense data) down to level 6 (philosophy), meaning that abstract knowledge about a topic will not change your sense perception. Just abstractly thinking about it will not help, but you can develop strategies to put yourself in situations (level 1) to challenge yourself and grow. For example, you can check an optical illusion (for example, the Hering illusion, see Chapter 5.3.2) with a ruler and conclude that the lines are indeed parallel and straight. But this abstract knowledge will not change your sense perception of the optical illusion. Similarly, anything you have learned in this book will not have an immediate effect on your self (or your self-perception). Instead, you first have to put this abstract knowledge into practice by creating new habits in order for your natural behavior to change. Knowledge needs to be integrated not just in your conceptual faculty, but in your entire brain by experiencing cause and effect.

Animals that possess only level 2 of consciousness might have a working memory, too, but they lack the flexibility to take the view of another animal (or person). They have to learn social behavior through trial and error. In animals with level 3 of consciousness, the brain machinery (the TPJ) is required to make the association. If the TPJ is damaged, the information that is unassociated is no longer available to further process in the neocortex through the working memory. It is like saying "Has the key." But without knowing *who* has the key, the information is useless. As we have seen with hemispatial neglect, something might still evoke a subconscious reaction even though it is not available in the neocortex: it might still be processed by other brain parts like the amygdala and thalamus. To actually make use of the working memory, we need of course a prefrontal cortex to update our models of the world and use those to suppress other brain parts (for example, resisting a candybar).

Just like the output of any other model, we are not always aware of the output of the model of attention. A moment ago, you were not likely focused on your toes. Now that I have mentioned it, that fact is brought from your body schema to your attention. Your brain is now focusing on that fact and there will be an entry about your toes in your working memory. Similarly, I could ask you the question "What are you currently looking at?" and the output of your model of attention suddenly gets pushed into the working memory: you are currently looking at words in a book. In that regard, it is important to note that we do *not* have a conscious experience of self all the time. We are not constantly reflecting on ourselves from an observer role. Most of the time, we are engaged in a situation and not very self-aware. When we are self-reflective, we simply ask ourselves repeatedly about our attention schema. This is just like someone else might ask us about how we perceive the world, who we are, what we are currently thinking, or what we are looking at. Similarly, we only become aware of what *others* are focusing on if we are actively focusing on their behavior. One could call it a "stream of consciousness" if you were to constantly listen to and ask yourself: "How am I doing?" or "What am I thinking at the moment?" We might experience a stream of consciousness when learning something new, especially about ourselves. Whenever we encounter surprising sense data from our environment, it comes into our attention because we need to update our models—which in turn is also something worthwhile to communicate to others to let them know what we plan or what we feel. Being aware of this stream of consciousness requires us to manipulate the working memory, meaning we need to have an awareness schema (level 5). This also allows us to imagine a future and past and to be able to play games like chess, which requires thinking ahead about possible steps our opponent will take and how we could respond.

Finally, philosophy and neurobiology (level 6) can help us to develop better habits, therapies, and even cultural norms (ethics). What stands out in the discussions of level of awareness is the fact that none of the items mentions *consciousness*. In that regard, it is best to look at consciousness as an *umbrella term* for all 6 levels mentioned above, combined, with different people and animals having different levels of consciousness. Consciousness also includes the subjective experience which we will discuss in the next section.

CONSCIOUSNESS · *Consciousness* is an umbrella term for the brain's abilities to process sense data, focus on something, be aware of something, have the ability to process high-level information (attention schema), be able to control awareness (awareness schema), and reflect on abstract information (philosophy and science).

In terms of the evolution of consciousness, we know that magpies, chimpanzees, elephants, and possibly dolphins hold funeral-like gatherings when one of their group dies. This implies understanding of the past and the future. Likewise, they can also solve complex tasks and use tools. In that regard, more research is necessary. The more we know about other animals with similar brain structures, the more we can uncover about ourselves and our own evolution. Consciousness is not something that suddenly came into existence, but lies on an evolutionary slope from the sponges' calcium signalling, the *Hydras'* processing of basic sense data, the arthropods' ability to focus, the reptilians' ability to be aware, the mammals' (and possibly birds') attention and awareness schema, and ultimately our ability to conduct neurobiological research and to think about philosophy.

6.5.4 Awareness Schema and Language

A different perspective on our awareness schema is to imagine how our experience of life would be if we did not have it. And it seems that people who have never learned a language presumably never learned to use their awareness schema. After all, if nobody (not even ourselves) asked who we are, what we want to become, and what we are thinking about, why would we feel inclined to use language or think about the past and future? Language could be our "mental limb" which we can "kick" to explore how our mind reacts—just like a baby kicks her feet to build her own body schema. This is supported by the fact that we are using the same parts of our brain to calculate limb positions and to use grammar. Without language, we would still have a working memory that allows other models to access high-level information, but we would be unable to use the information about what we are currently doing or thinking to affect our behavior in the next moment. We would be unable to write instructions into the working memory, our brain would have not been able to build an awareness schema to perceive the past, present, and future. It might even have only a rudimentary attention schema, as a necessary requirement of accessing our working memory is to have a concept of "self" and "other." Only by putting yourself in relation to your environment can you form a concept of self (and other). The requirement for this is language. As a glimpse into how it might feel to experience the world without an awareness schema, this description from Helen Keller is revealing. She became blind and deaf due to meningitis when she was only 19 months old; she put her experience before she learned a language this way:

> Before my teacher came to me, I did not know that I am. I lived in a world that was a no-world. I cannot hope to describe adequately that unconscious, yet conscious time of nothingness. I did not know that I knew aught, or that I lived or acted or desired. I had neither will nor intellect. I was carried along to objects and acts by a certain blind natural impetus. I had a mind which caused me to feel anger, satisfaction, desire. These two facts led those about me to suppose that I willed and thought. I can remember all this, not because I knew that it was so, but because I have tactual memory. It enables me to remember that I never contracted my forehead in the act of thinking. I never viewed anything beforehand or chose it. I also recall tactually the fact that never in a start of the body or a heart-beat did I feel that I loved or cared for anything. My inner life, then, was a blank without past, present, or future, without hope or anticipation, without wonder or joy or faith.

—Helen Keller, *The World I Live In and Optimism: A Collection of Essays*

6.5.5 The Subjective Experience

While we have explored consciousness and its origin, looking at the diagram (Figure 6.30), we still find no evidence for our subjective experience. While the model explains how something can *enter* subjective experience and how we can think about alternative scenarios, it does not yet explain the subjective experience itself. In philosophy, this question is also called the *hard problem of consciousness*. If we dismiss the Cartesian theater with a homunculus sitting in our head, then who or what is experiencing what we see, feel, and hear? While we have explained how the data flows through the head with the loop of consciousness, a loop does not magically create a subjective experience.

> **HARD PROBLEM OF CONSCIOUSNESS** · The *hard problem of consciousness* asks the question where the subjective experience of consciousness comes from. It is "hard" as there are no known ways of detecting this experience objectively without relying on the subjective claims of an individual (or ourselves).

The problem is that the only source that claims that we have an inner experience seems to be us. Even if we assume that our own subjective experience is "real," the only possibility to inquire about the subjective experience of other people seems to be to ask them and trust that they are telling the truth. Or they could be the victim of a delusion and tell us they have a subjective experience when they do not. After all, we could program a robot to do nothing but answer "I have a subjective experience" and explain in great detail what it is (supposedly) experiencing. In philosophy, such beings without a subjective experience (or more generally, without consciousness) are also called "philosophical zombies."

> **PHILOSOPHICAL ZOMBIE** · A *philosophical zombie* is a hypothetical being that acts exactly like a human but lacks the inner conscious experience. The existence of a philosophical zombie is used as an argument against consciousness being a mechanistic

> process: if we are but a mechanistic machine, why would we need a subjective, conscious experience? This argument is addressed by pointing out the necessary evolutionary advantage of having a conscious experience, which is the ability to focus and to explain oneself to others.

Ultimately, the question is if a being that somehow does not have a subjective experience can be distinguished from a being that has one. To examine this, we need to show the evolutionary advantage that subjective experience provides. We have to draw a diagram with one box labeled "subjective experience" and show how this architecture will be able to solve certain problems better than one without subjective experience.

As we have explained all other properties of our subjective experience (attention, awareness, and awareness schema), what is actually left for the subjective experience to do? *What are the specific properties of subjective experience that are not yet covered by awareness and attention?*

To answer this question, let us take everything we have learned about consciousness and program it into a robot. Then we can discuss whether or not the robot has a subjective experience. One thing is certainly true: if the subjective experience is supernatural, this constructed computer will not have it.

The robot will have the same basic setup as discussed earlier in Figure 6.30. Sense data from the camera eyes arrive at the robot's visual cortex, are analyzed in its dorsal and ventral streams, and then combined (together with an identification of *who* is having the perception) in its TPJ. The results are put into its working memory. At any time, the prefrontal cortex of the robot can access the working memory, which contains images, sounds, and abstract information.

You could ask the robot: "Do you *see* the information in your working memory?" The robot's temporal lobe will analyze the data, and then it will be processed in the robot's TPJ as well. The robot can properly interpret the question, recognizing that it should reply whether or not it sees something. Regarding working memory, though, it will inquire what exactly you mean with "working memory" in the context of "seeing."

You reply: "Let us first agree on what you mean by 'seeing.'" You put a red apple on a table. "Here, that is a red apple. If you can identify it as a red apple, you are seeing the apple."

The robot will connect the idea of "seeing" with information passing through the visual cortex to the working memory. "OK, I am seeing the apple." What is happening is that the robot is using its attention schema—meaning it can access the working memory, but it does not know that it is accessing its working memory. Just like we are using our body schema but cannot explain what specific muscles we are using when we are, for example, walking.

You reply: "OK, close your eyes and think about an apple. Do you see the apple?"

The question will make the robot deactivate its cameras and activate its memory of an apple, which in turn activates the occipital lobe. Now similar information flows through the robot's brain as if it were looking at a red apple. Ultimately, the information about the red apple will appear in the working memory and the robot will say "Yes, I see the apple."

You reply: "But do you *actually* have a subjective experience of the apple?"

Assuming the robot is philosophically well-versed, it might answer: "What is the difference between seeing, hearing, feeling, touching, tasting, or smelling something and having a subjective experience of it?"

You might respond "Well, when I smell roses, I smell roses. I do not read from my working memory the abstract information that I have just smelled roses. There is more to experiencing something than sense data."

Indeed, our subjective experience seems to be connected to our sense perception. And indeed, with its interpretation of "seeing," the robot *sees* the apple. The challenge we have with this answer is that we usually attribute a much richer experience to our perception of the world than just reading a line like "I see a red apple" from our working memory. Even when accounting for subsequent memories that sense data evoke, our subjective experience still seems to be "richer" than just being abstractly aware of the sense data.

6.5.6 The Limitations that Make Us Human

With our awareness schema, we have the misconception that we are able to be aware of all the details we are perceiving, when in fact, we we can focus only on a small number of details at any given moment. Again, it is like the experience of our body. We think we know what we do when we move the arm, while in reality, our brain orchestrates dozens of muscles to, for example, grab a glass of water.

Outside of our focus, the level of detail decreases rapidly. When focusing on the head of a person in a painting, the rest of the painting becomes a "room" or "garden." When looking at the flower in the picture, the individual facial features and the facial expression becomes "face." Ultimately, we should add the *source of experience* to our working memory as well. A visually imagined apple is something else than if we heard the word "apple." While pointing to the same thing, the experience is different.

Ourselves, for everything we do not focus on, we drop the measurements and just remember the concept, while a robot might save all the information. The way our brain is optimized, we prefer comparisons ("This smell reminds me of ...") to absolutes ("The molecular composition of this odor is ..."). That means we *are* aware that we have more information. Just because we are aware of something does not necessarily mean we are able to express it. This difference is a significant part of what we call the unconscious or the subjective experience.

Could we create a human-like subjective experience in the robot by limiting its memory?

Following our example, the robot could drop the measurements and process concepts instead of raw sense data. Suddenly, *he* might seem much more human-like. He will see a small table in a large room, not a 2 by 1 meter wooden table in a 20 by 10 meter room. And when being asked about the table, he can focus in, remove the information of the 20 by 10 meter room and instead load information about the table's legs, texture, color, etc., and that there is an apple on the table.

When asked about the apple, he will again reduce his focus and move up the tree to more general concepts, removing details about the table, and adding information about the apple (red, juicy, small, ...). Hence, we can now ask him:

"Robot, reduce the size of your memory and tell me what you see in this picture."

"Reducing my memory that describes the world. OK, done. I am seeing a red apple on a table."

But he is wrong. Below the table, there is also a pair of shoes and they are in his field of view. Let us ask him:

"What about the shoes below the table?"

"Right, now that I am focusing my attention on things below the table, I also see the green shoes. I was not aware that there were shoes there. Previously, that information did not appear in my limited working memory."

Having only a limited amount of memory, we need to assume that sense data in our brain is not a two-dimensional image or sound recording, but instead represented as a tree-like structure on which we can zoom in or zoom out. We can explore any detail by traversing up the tree to the leaves, or generalize the sense data by moving towards the trunk. Depending on the size of our working memory, we can look at only a limited number of leaves at the same time. The fact that our attention schema can move to any leaf we want at any time provides the illusion that we are looking at the full picture. We *are* aware that we *could* examine any part of our working memory in more detail. Evolution has balanced the size and energy costs of the working memory with the benefit. For most tasks, it is enough to focus on one thing in great detail while keeping a rough idea of the rest. For example, you can examine the texture of an apple while also still being aware that you are sitting in a room.

Our thoughts are like a sea with waves where only the largest ones translate into discrete concepts we can communicate.

Even for seemingly simple sense data like the spoken word, we do not just hear the word and then conceptualize it. Spoken words convey emotions, they tell us something about the speaker, his origin (through his dialect), maybe even his health and occupation. All that context information is processed in the brain and is usually available only in your subconscious. Maybe the words themselves remind you of memories. Some words might conjure images in your mind, which again might evoke an emotional reaction.

Brains are not digital computers. Neighboring brain parts can influence each other, despite not being thematically related. For example, some people experience synesthesia, which is a condition in which sense data flows over to other brain parts. They might experience numbers, words, or music also as colors or shapes and vice versa. Even most people without synesthesia associate certain speech sounds with round or spiky shapes (the "bouba-kiki effect") which points to information overflowing from the temporal lobe to the parietal lobe and the other way around.

Yet another point is that our brain is not just the neocortex. Instead, it is a construct of overlapping responsibilities, usually multiple brain parts dealing in different levels of refinement with similar issues. Perhaps the more primitive systems in our mid- or hindbrain could act in a similar way as the system of attention and awareness in our neocortex that interprets our sense data. In that regard, what we have skipped in our discussion so far is whether or not our subconscious is actually part of our subjective experience. After all, we know from people with blindsight or hemispatial neglect that while they are not consciously aware of what they are seeing, experiments have shown that other parts of the brain produce a "gut feeling." In that regard, we can assume that *any* perception produces a gut feeling, even though our conscious experience usually overlaps that feeling. The idea that the subjective perception is equal to reading the working memory ignores the fact that every perception is accompanied by a subconscious feeling we cannot put directly into words. We can be sure that the robot does not have this subconscious feeling—we have never programmed it to have one.

Finally, a lot of information is lost when trying to express what we have in one type of working memory (for example, our visuo-spatial scratchpad) with another (for example, the phonological loop). What we experience as visual sense data might not be fully describable by words, and thus we attribute a richer experience to sense perceptions that we cannot describe as abstract concepts.

In summary, with the attention schema, we have seen that even a simple robot can have a subjective experience, although *we* would not call it that. For us, beyond the abstract information coming from the working memory, many more (partly subconscious) signals go through our heads which we cannot always describe in words. Our brain is not designed with a clean architecture. Instead, it has parts with overlapping or even duplicate functions.

If we removed our genetic heritage and cleaned up the brain's architecture to create discrete brain parts, our experience of the world might be similar to that of a robot—reading entries from a limited working memory and organizing sense data according to the attention schema in a knowledge tree. Now, does that mean that the way we experience the world is but a side-effect of evolution? Maybe to a small degree, but we should not forget that the sum of all our parts allowed our ancestors to evolve to become humans. Sometimes, a purely rational choice based on evidence might not have been the choice that led to our survival. As such, how we experience the world is an echo of nearly a billion years of history—something missing in the way we constructed the robot.

6.6 Building a Conscious Mind

With what we have learned about consciousness, we will re-examine the evolution of attention discussed in Chapter 5.1, and expand it with the attention schema and awareness schema. Instead of focusing on animals of the past, we will focus on a single example of a co-evolving predator-prey system. Imagine a group of tigers and a number of small apes. The tigers are hunting the apes, and the apes are trying to evade the tigers while at the same time trying to find fruit. They are living in a forest where trees provide cover, and where berry bushes provide a (limited) source of food. The forest also contains berry bushes with poisonous berries, and some of the non-poisonous berries are not yet ripe and thus not edible. Over generations, the tiger population evolves better ways to catch prey, while the ape population focuses on improving her abilities to forage, and evade predators. In the first part of this section (unconscious strategies), we will build up the cognitive abilities of the apes (and tigers) step by step using elements we have learned in Chapter 5. In the second part, we will enhance the behavior using the attention and awareness schema.

6.6.1 Unconscious Strategies

For the first part of this thought experiment, we will look at strategies using basic parts of the brain like the sense organs, the amygdala, and the hippocampus. Initially, we use a "dumb" tiger as a predator. It will simply run toward an ape when in sight, but will wander around randomly at other times. For the tiger to be successful, all it needs is to detect movement, so a stereoscopic view (and all the mental machinery necessary) is not (yet) necessary for the tiger.

As the tiger could run in any direction, it needs to have something that leads it to prefer one action over the other. To train its own neural network, it needs a reward system (supervised learning) which includes the amygdala. In the beginning, the tiger might wander around aimlessly. But once it has associated meeting (hunting) an ape with food, it will seek out more potential prey (more apes) in the future. It will have associated apes with a positive emotion. To better track the ape, it might even evolve a *superior colliculus*, which improves eye movements related to moving targets. Now, as the tiger begins to learn to hunt the apes, the apes themselves might evolve better strategies to forage food and evade the tiger. Given how the brain evolved over time, the following list can be seen as one strategy layered over the other, with more brain complexity required for each strategy.

Random. The most basic strategy for an ape would be to just act randomly. This could be walking around, staying close to berry bushes, trying to eat berries, and (lacking sense data) might even try to nibble on trees, tigers, or other apes. That strategy is of course far from optimal because it does not increase the chances of finding berries, getting away from a tiger, or avoiding poisoning from certain berries.

Pain and pleasure. We give the apes a sense of touch, and a sense of taste. If the sense of taste detects toxins, or the sense of touch physical pain, the ape might refrain from continuing to eat something (for example, trees or poisonous berries). Vice versa, if the sense of taste detects nutritious food, she will eat the berries more eagerly.

Tracking inner state. Eating more food than the ape needs unnecessarily increases the chances to get food poisoning, and will deplete food sources. Tracking the current internal state to generate a feeling of hunger or satisfaction can regulate the pain and pleasure mechanism. This already is a primitive model of the body to control food intake: instead of eating everything in sight and then going hungry until a new source of food is found, consumption is distributed over time.

Sense of smell. Adding a sense of smell (the olfactory system) allows an ape to decide in advance whether food is good or not. For this to work, the apes need an amygdala that connects sense data with the pain and pleasure mechanism. Smells associated with a positive experience (nutritious food) are evaluated as positive; smells associated with a negative experience (from eating trees or poisonous berries) as negative. An ape might even learn to associate the smell of the tiger as something dangerous (assuming the ape survives an encounter) and end up not trying to eat the tiger in the future.

Sense of sight. Next, the apes could evolve a basic sense of sight with eyes, their retinas, and a superior colliculus. Connecting the sense data from the eyes to the amygdala, an ape could learn to associate the visual information from berries, tigers, and trees to decide from afar and without needing to smell them to determine whether or not they are dangerous. This is of course only a small advantage as the ape has no sense of direction and all she can do so far is to stay and eat or move around randomly.

Directed movement. Using back propagation with the amygdala as the supervisor for supervised learning, apes can learn more complex movements than chewing (or spiting out) food. The apes will still move around randomly if no entity is in sight. They will either approach or move away from the strongest signal. For an ape as we know it, the challenge is to not only use the visual information, but also the information about which direction the head and the eyes are turned. Our apes do not have this *model* of their body (yet), so we just assume that neither the eyes nor the head can turn. In that regard, "seeing" always means "seeing in front of us."

Turn toward movement. The primitive amygdala we gave the apes does not recognize particular entities (like the visual cortex does). Instead, it takes all the incoming sense data and decides whether or not it fits a previous experience. Hence, a berry bush in the open field does not provoke the same degree of response from the amygdala as a berry bush in the forest. While there are some similarities, the ape needs to learn those situations separately. In particular, this means that the ape cannot analyze the visual field to decide in which direction to turn. There is no understanding by the ape that something on the left side of the visual field can be seen better by turning her head to the left. Instead, that movement would have to be initiated by a genetically programmed reflex. Then the ape can turn toward a movement, identify the possible threat, and then use the amygdala to run away from it.

Refined senses. Next, the apes might evolve a sense of hearing. This improves the apes' ability to detect a nearby tiger. Instead of being able to look only forward, they will have a 360-degree perception of possible dangers nearby. When an ape hears something from a particular direction, she turns toward that direction and her amygdala can react to the new sense data she sees. For the actual comparison, the thalamus is required. It integrates sense data from different sources (in this case, the left and right ear) into abstract information (direction). The thalamus could also create correlations between visual sense data and auditory sense data to strengthen the signal (something is moving and it makes noise, for example, a tiger) or weaken the signal (something is moving and it makes no noise, for example, a cloud).

Filtering sense data. Beyond combining sense data, the thalamus and the connected cortices (auditory cortex, visual cortex, and olfactory cortex) preprocess sense data to arrive in the amygdala with less noise. Noise reduction is important for the amygdala as it learns based on absolute sense data (e.g., the image of a specific apple) and not abstract concepts (e.g., the concept "apple"). For example, a tiger in the shade should be handled similarly to a tiger in the sun. Likewise, just because the ape hears some background noise, the roar of a tiger should elicit a similar response as a roar at night when it is otherwise quiet. For filtering, the thalamus can also combine the information from different senses. For example, it could use visual information like the location of the origin of a sound to improve its auditory filtering. Vice versa, being able to locate the source of a sound with the ears might help to locate the source of a movement with the eyes.

Resolving conflicts. At any single moment, the ape can move only in a single direction. But different thought patterns might try to take control of the muscles at the same time, with one part of the brain wanting to go left while the other wants to go right. For example, in which direction should the ape run when she sees two tigers approaching from two different directions? In the worst case, the ape might end up doing nothing because she cannot decide. To arbitrate such conflicts, the basal ganglia can at least lay out basic rules of how decisions are made and then clearly communicate that decision to the muscles. One such rule of the basal ganglia could be a simple majority rule: execute the action dictated by the strongest signal and suppress the other signal. For the ape, it is better to run into one direction at her maximum speed than to run half-heartedly because she is worrying about the direction she has not taken.

Keep doing things. Next, the apes could evolve a hormonal system (hypothalamus, pituitary gland) to temporarily remember positive and negative experiences. For example, when the ape sees a tiger, the amygdala might induce fear by releasing a hormone through the hypothalamus. As long as that hormone remains in the blood, the ape's organism could adapt its body chemistry to either run faster (and keep running in one direction, even if the tiger is no longer in sight) or stay frozen in place so that the tiger does not see the ape. Similarly, it could cause an increase in appetite so that the ape stays near a berry bush to eat all the berries before moving on.

Remember locations. Adding the hippocampus to store memories of locations will help the ape to find her way to nearby berries more quickly. A certain formation of trees might make the ape remember that she has found a lot of berries nearby. By associating places with each other, the ape could remember a path toward a field of berries. This works in a similar way as backpropagating for neurons: the ape can associate a place positively if she encountered it before reaching an already positively associated place. For example, a tree-lined path might lead to a group of trees which in turn lead to a group of berry bushes. It can also help the ape to find a path around an obstacle such as a group of trees.

Working memory. Sense data is fleeting, so at least in the short-term, it would be good to keep recently perceived sense data present. For example, if the ape heard something on her left and turned her head, the sound might no longer be audible. Still, knowing her previous location will help the ape to visually identify the location more quickly. With the phonological loop and the visuo-spatial scratch-pad, the ape can correlate past with current sense data. Both mechanisms form the working memory and send sense data into a loop between the auditory cortex and the thalamus or the visual cortex and the thalamus. The model the brain uses to access the working memory is the attention schema.

Empathy. If the ape sees another ape throwing up food, it might not be a good idea to eat from the same berry bush as it could be poisonous. As she has no understanding of how berry bushes work, the mechanism would need to evolve slowly over time as a genetically programmed instinct. A mechanism could evolve to recognize basic facial expressions that influence the pain and pleasure mechanism and the tracking of the internal state. As if she had the experience herself, she would be repulsed either by the specific berry bush or (with the help of the hippocampus) the area itself.

Learn the environment. The hippocampus can also store a timer for the ape to return once the berries have grown back. The ape might mentally note to not visit recently harvested berry bushes soon. If there are no known (and ripe) berry bushes nearby, the hippocampus could specifically favor exploring places the ape has not been before, judging already visited places as slightly negative. It could use similar pathways that are used for tracking the internal state to determine hunger. Instead of having just a one-dimensional hunger feeling, the hippocampus generates a feeling of hunger for particular places. Just like we might prefer going to a particular restaurant to order our favorite dish that satisfies our taste (and nutritional needs), the ape would refrain from visiting a place with depleted berry bushes in favor of another place with an abundance of berries.

Flexible eyes and head. Making the head and eyes flexible allows the ape to save energy by turning just the eyes or head instead of the whole body. This requires an extension of the model of the internal state of the ape (that, so far, tracked only hunger) with limb positions by combining sensory information from the skin and internal sensors from the muscles in the somatosensory cortex. By correlating visual sense data with head movements, the position of the eyes and head relative to the environment and body can be calculated. For example, when the ape looks behind herself and sees a tiger, the tiger is in front of her eyes. It is not in front of her body, though. She first has to translate her turned head and eyes to the relative direction of her body. This forms the dorsal stream of sense data between the visual cortex and the somatosensory cortex.

Smooth movements. At this point, the ape still uses instinctive movements to turn toward movement or noise. For example, the ape sees something moving on the right side of her visual field. She could now turn the head and eyes a little to the right, check her visual field again, turn again, and so on, until the interesting part of her visual perception is in the center of her visual field (with maximum definition). However, this going back and forth takes a while and—as with the simple thermostat—might cause under- or over-shooting the target. To turn the head faster (and more accurately), the model of the head and eyes can be used to predict how far the head needs to be turned for the target to be in the center of the vision. The ape faces the same problem when using her limbs. To move faster (and with fewer accidents), the ape has to coordinate the movement of her legs and arms. She has to predict what movement leads to which positioning of her limbs. Once the basics of this prediction system are in place, the evolutionary race with the tiger (with the ape getting more agile with each generation) is on. Over time, the motor cortex, somatosensory cortex, and nearby areas of the supramarginal gyrus and angular gyrus evolve.

Motor programs. Enhancing the control of muscular movements, the brain can combine movements into motor programs. These motor programs are like a sheet of music, indicating starts and stops for different instruments (muscles). For this, the ape needs enhanced basal ganglia to not only arbitrate currently conflicting thought patterns, but also to coordinate thought patterns over time. The basal ganglia can achieve this by suppressing or promoting movements in a certain sequence governed by the motor program. A simple example for such a motor program would be turning the head. This could replace the movements the ape previously had to learn separately for each sense data combination. Turning to the right because the ape has heard something coming from the right could use the same motor program as turning to the right because the ape has seen something moving on the right side of her visual field.

Enhanced empathy. Combining the mechanisms for analyzing the face to detect whether or not another ape has eaten poisonous berries, the ape could reuse her supramarginal gyrus and angular gyrus not only to read her own limbs to build a body schema, but also to interpret the posture of *other* apes. This way, she could indirectly access other apes' senses. For example, she could approach them if she sees them munching on berries, or run away if she deduces that they must have seen a tiger.

Form groups. When apes form foraging groups, they have much better chances than individual apes to detect an approaching tiger. The preference to form a group could easily evolve by being able to recognize apes and judging them as "positive." This could evolve genetically or through learned experiences during childhood.

6.6.2 Conscious Strategies

While the strategies discussed in the previous section do not require awareness, they do show relatively complex behavior. The main challenge of this approach is that the ape with this basic setup would be able to use only supervised learning. This means that discovering her environment requires either an instinct, or through slow and possibly dangerous experiences in the real world. While it works well with rather static entities like regrowing plants and with reactive behavior (flight, fight, or fawn), it is problematic in dynamic environments or with situations with a lot of variables. This is no different than the thermostat we discussed in Chapter 6.4 which also had problems with, for example, changing locations or seasons. Similarly, trying to communicate to other apes where the tiger will probably show up requires higher brain functions. Here we will examine how we can get there step by step, again with strategies that layer on top of each other.

Unsupervised (long-term) learning. Let us add a more diverse flora to our imagined world. The properties of a single type of plant can be learned very quickly, even with supervised learning. But if an ape wants to learn about multiple types of plants, and wants to recognize different sizes and development stages of the plant, she needs unsupervised learning. What the ape needs is a link between the concept of, for example, a berry bush, and the (one-time) experience the ape had consuming berries. For the brain to handle concepts and properties, apes need to evolve a declarative long-term memory, and a process to transfer the sense data to the memory (dreaming!). To connect one memory with other memories, the ape can use her hippocampus' ability to connect different locations with each other. What the apes now have to evolve is a connection between the declarative memory and the amygdala. Now, instead of the amygdala having to learn each entity separately, sense data are categorized first. For example, different (even yet unknown) types of plants could be categorized as "nutritious" and would all evoke a similar reaction from the amygdala. If she finds a plant that is poisonous after all, she would have to reevaluate her categorization of plants.

Instinctive communication. By adding the ability to make sounds, the ape could communicate danger. Using unsupervised learning to categorize calls and create mappings between observations and calls, this can lead to a basic language. For example, the ape could observe her parents making a certain call when finding berries. The temporal lobe near the ears and could serve both as a way to encode as well as decode individual calls. A loud shriek might indicate that a tiger is seen nearby, while a longer low call might be associated with a food source. This communication would still be unconscious and a direct result of, for example, the amygdala's emotional response (danger, food, etc.).

Combine information. Information about the what and where (processed by the dorsal and ventral stream in the parietal and temporal lobe) could be combined in a new brain part (the TPJ) to build an abstract description of the world. This enhances the *working memory* to provide the information to other parts of the brain. In a first step, this could improve signal processing combining the phonological loop and visuo-spatial scratchpad. We see its workings with auditory and optical illusions where both bottom-up processing as well as top-down processing happen. If the ape knows what she is looking for, her brain can use this information to process the image more effectively. For example, berry bushes grow on the ground, not high up in the tree.

Tactics and object permanence. The ape is already using models to track her inner state, to have smoother control over her movements, and to interpret the state of other apes or the tigers. Adding a prefrontal cortex that constantly reads the working memory to create an internal model of the entities in the environment allows smoother tracking of entities. Instead of being surprised when a tiger disappears behind a tree only to be surprised again when it reappears, the prefrontal cortex can mentally track its position. This allows an ape to "keep in mind" that there is a tiger behind the tree. This hidden tiger could be used by the prefrontal cortex to continue to project fear into the amygdala, causing the ape to be more cautious or even leaving the area—even though the tiger is no longer visible.

Episodic buffer. Connecting the hippocampus to the working memory allows the ape to distinguish events in their temporal sequence. This helps with interpreting what is happening. The sound of a crouching tiger means a tiger is nearby, the sound of a crying ape means that an ape is injured. Both in combination might point to a successful hunt by a tiger.

Planning. The ape could be faced with different options to forage. For example, should she take the risk of walking all the way to the large group of berry bushes (higher risk but also higher reward) or take no chances and just forage the berry bush nearby? While choosing the closest berry bush is an effective strategy, it might not be the most efficient. Depending on the risk associated with a particular area, as well as the distance to the area, the ape might prefer one option over another. As the relative distances change all the time when the ape moves, the amygdala's positive or negative evaluation of a place is not sufficient for the decision. That risk could be calculated by the prefrontal cortex and used to modulate (suppress) the amygdala.

Pathfinding. With access to a model that contains (visible and hidden) entities of the environment, the ape could calculate in advance the shortest path to a location. For example, the ape might know that a bush of berries is behind a hedgerow. Instead of walking toward the hedge and then trying to find a way around it, the ape could calculate in advance the shortest path. To accomplish this task, the prefrontal cortex might need a connection to the basal ganglia and actively suppress decisions to walk outside the optimal path.

Classifying berry bushes. While the hippocampus already uses a temporal component to not visit berry bushes that were only recently harvested, using a model of berries might increase the harvest. There is a specific time window in which harvesting a berry bush is optimal. Too early and the fruits have not sufficiently grown, too late and the fruits are probably already harvested by another ape or have fallen to the ground. This requires classifying the berry bush according to its development phase and suppressing the urge to visit the berry bushes when the model says that they do not carry any ripe berries.

Language. Replacing supervised learning with unsupervised learning and using concepts from her temporal lobe instead of instinctive behavior ruled by the amygdala, the ape could start to communicate more effectively. For this, a connection is needed between her temporal lobe and frontal lobe with the motor cortex (in humans, between the Wernicke's area and the Broca's area). For example, she could specify what type of danger she is seeing, rather than having to learn each call separately for each situation.

Advanced tracking of the inner state. The ape's ability to find the shortest path to a goal can be improved by combining it with the internal state tracked in the primary somatosensory cortex. Tracking also caloric intake, energy, sleeping cycles, and walking speed, the ape might decide not to simply seek the berry bushes with the best reward/risk factor, but instead to strategize using all known factors. For example, the prefrontal cortex could activate the amygdala to increase hunger, which the parietal lobe would translate into eating a lavish meal before the journey, or making a detour to gather berries on the way to a larger field of berries.

Focus and intelligence. Just being able to plan ahead and suppress particular movements might not be enough to actually follow through with a plan. Given that the brain parts compete with each other, a stronger emotion might overrule the influence of the prefrontal cortex. For example, we can plan exercise and diet but ultimately fail when being confronted with dessert. Instead of trying to overpower the amygdala directly with the neurons in the prefrontal cortex, the apes could evolve a way to recruit other parts of the brain to influence the working memory (the awareness schema). Using the new connection between the motor cortex and the temporal lobe, the ape could use it in the opposite direction and activate concepts by thinking about them. The act of thinking would use the same pathway as a physical action. But instead of a motor action at the end, the motor cortex activates the temporal lobe. Hence, "thinking" feels the same way as executing a physical action. Standing in front of an apple the ape cannot immediately reach, she might use her "mental arm" to think about the apple and recruit the rest of her brain to devise a plan. This activation then goes the same route as an activation by the visual cortex, namely into the working memory. From there, it goes back to the thalamus and back into the neural competition and then possibly through the basal ganglia again into the working memory. With this loop, the ape can recruit multiple parts of the brain to, for example, overpower the signal from the amygdala and continue with a more important goal.

Imagining a future or past. With each of these cognitive loops, the ape could also build up a scenario in her mind. Instead of just using the loops to suppress the amygdala, the ape could think one step ahead, have other parts of the brain process this future scenario, and use the result to calculate the next step. For example, the prefrontal cortex has no machinery to imagine three-dimensional structures. But the visual cortex and parietal lobe do, so mentally rotating an object requires the interaction between the prefrontal cortex and the parietal lobe and visual cortex. Helpful in this endeavor is also the episodic memory, which puts thoughts into a sequence. By imagining a future scenario, the ape can better plan ahead. For example, she could imagine the next day and gather more berries than she could eat now and store them for later.

Understand causality and build models. Once the ape can track time and imagine herself in the future, she can also track how properties of other things change. This means that the ape can now conceptualize processes and causality. This allows the ape to observe other apes and understand that the presence of apes diminishes the available berries over time (well, they eat them). This model can be used to enhance the ape's strategy to evade places with a lot of other apes as they might pick the berries before the ape can reach the berry bush. Similarly, the ape can build models of the tiger's behavior (not just its current state) and how the tiger hunts apes. She might see the tiger less as a threat if there are other apes nearby, as the tiger can hunt only one ape at a time. To make more informed decisions, the ape could also take note of how fast the tiger can run. For example, the ape could observe that a tiger that just has eaten is not as fast, so the ape could come closer to the tiger. Understanding object permanence allows tracking possible positions of other apes and the tiger. Furthermore, the ape could observe whether or not the tiger has noticed her and sneak away. The ape could also suppress her urge to call out that there is a tiger if the tiger has not noticed her yet. Similarly, the ape could observe whether or not another ape has noticed her, and decide to steal some berries.

Ultimately, being able to look into the future forms the foundation for goals and empathy. Instead of, for example, just assuming tigers are bad because they evoke a negative reaction in her amygdala, she could now imagine what would happen if she got injured or even killed by the tiger. Besides the pain involved, she might be thinking of her family who would starve if she did not return with berries. Her fear can then become *rational* as she would be acting on her values and on thought instead of just instinct. Similarly, she could refrain from eating berries if relatives are nearby or even stand in the way of the tiger to sacrifice herself to protect her relatives.

Tool use. Besides pure logic and causality, she could also recruit the SMG and AG to think about complex mechanical problems and consider tools as an extension of her arm. It requires three-dimensional thinking about how, for example, her arm, a stick, and the target relate to each other.

Gaze following. Enhancing her abilities to analyze the way eyes of other apes move, she can more quickly detect possible danger. She can also refine her theory of mind of other apes, the model of their inner worlds. By knowing what they pay attention to, she can predict more accurately what they will do. The more the ape interacts with other apes, the more she will know what other apes know. She can deduce what they are seeing, hearing, or thinking. This is a prerequisite for better communication. For example, instead of always calling out when she sees a tiger, the ape could first check whether other apes are seeing the tiger, too. If they do, calling out would just draw attention from the tiger to the apes. Likewise, by observing the tiger, the ape can decide whether or not the tiger has already seen her.

Deception. She could also use her insight into what other apes know to deceive them. For example, by making the "tiger call," she might scare other apes away so that she has all the berry bushes for herself. Apes could also develop an evaluation for each ape of the tribe. Tracking outcomes of interactions, they might find out who is helping them out and who is deceiving them. Apes could demonstrate their relationships with each other openly to others to demonstrate to all the apes present whether or not an individual ape is trustworthy. The ape could then reflect on how trustworthy other apes judge her. Similarly, the apes could develop a model of a culture that reflects what the tribe allows and what is not allowed, possibly keeping deceptive behavior in check. Having a network of trust, the apes could show more complex behavior following strategies like "I scratch your back, you scratch mine." Berries could be used as presents to improve one's standing in society. Especially strong apes might have a better standing as they might scare the tiger away if it is not particularly hungry.

Advanced language. Combining her abilities to think about alternative scenarios and about causality, she can learn to form more complex statements and grammar. This helps with planning a defensive strategy against the tigers. Together with tools, the apes might even start coordinating an attack on a tiger. Better language capabilities also help to form better relationships by allowing the apes to negotiate and form contracts instead of just relying on their feelings toward members of the tribe.

Explaining behavior. Explaining the reasons for her behavior helps other apes to make better predictions about the ape's future behavior. For example, she might decide not to share her berries with another ape. By verbally explaining that she needs the berries for her starving children at home, the other ape might not value her less. '

Writing and the scientific method. At this point, the natural evolution of the apes gets interrupted. With the development of language and tool use, apes will be able to write down their experiences to teach later generations or apes that are far away. This increases the apes' collective intelligence. The apes will be able to change their environment faster than their biology will be able to keep up. The apes will be able to think about how mental processes work, and will start **philosophizing about how they are able to reflect**. Maybe the apes will **start studying neuroscience** to discover how their inner cognitive reflection works. This way, the apes could counteract mental impairments or maladaptations to a changing environment. Knowledge of psychology and neuroscience can help the apes to understand and better process subconscious mental processes and emotions.

And here you are.

The Book Series
Philosophy for Heroes

She said, "I will go no farther." "There is no choice.
We can only go on." The magician said again. "We
can only go on."

—Peter S. Beagle, *The Last Unicorn*

The book series continues! Head over to our shop for more from this series: https://www.lode.de/shop.

Part I: Knowledge. *Philosophy for Heroes: Knowledge* dealt with the foundations of knowledge, philosophy and language. It also discussed how society defines a "hero" and how basic skills, such as language and mathematics, influence our thinking.

Part II: Continuum. *Philosophy for Heroes: Continuum* looked at the gradual transitions from one condition to the next. It dealt with questions of the origin of the universe and life, and introduced us to evolution and how it relates to our creativity.

Part III: Act. *Philosophy for Heroes: Act* discusses the materialistic foundation of consciousness. Knowing the evolution of our brain, how we compare to chimpanzees, and how we can explain our conscious experience helps us to be better understand ourselves.

Part IV: Persona. Being a hero requires not only courage and knowledge but also independence and consistency. *Philosophy for Heroes: Persona* sets the reader's mind free from harmful manipulation by others. It discusses concepts like free will and how ethics and psychology can help us to be independent.

Part V: Epos. The final book in the series, *Philosophy for Heroes: Persona*, examines the influence of the most powerful tool of a leader, the *story*. Is the age-old conflict between "good" and "evil" necessary? Do heroes need "dragons"? What can we learn from the ancient stories of religion? How can we use our language for good? How can our own lives become a story, an *epos*?

Recommended Reading

Calvin, William H. (1996). *The Cerebral Code: Thinking a Thought in the Mosaics of the Mind*. A Bradford Book. ISBN: 978-0262531542.

— (2007). *A Brief History of the Mind: From Apes to Intellect and Beyond*. Oxford University Press. ISBN: 978-01-9518-248-4.

Dawkins, Richard (1990). *The Selfish Gene*. Oxford University Press. ISBN: 978-01-9286-092-7.

— (1997). *Climbing Mount Improbable*. W. W. Norton and Company. ISBN: 978-0393316827.

Dennett, Daniel (1993). *Consciousness Explained*. Penguin. ISBN: 978-0140128673.

Graziano, Michael (2015a). *Consciousness and the Social Brain*. eng. First Edition. New York: Oxford University Press. ISBN: 978-0190263195.

— (2019). *Rethinking Consciousness - A Scientific Theory of Subjective Experience*. English. W. W. Norton & Company. ISBN: 978-0393652611.

Lode, Clemens (2016). *Philosophy for Heroes: Knowledge*. Clemens Lode Verlag e.K. ISBN: 978-39-4558-621-1.

— (2018a). *Philosophy for Heroes: Continuum*. Clemens Lode Verlag e.K. ISBN: 978-39-4558-622-8.

— (2018b). *Writing Better Books the Agile Way*. Clemens Lode Verlag e.K. ISBN: 978-3-945586-83-9.

— (2019). *Better Books with LaTeX the Agile Way*. Clemens Lode Verlag e.K. ISBN: 978-3-945586-48-8.

— (2020). *Philosophy for Heroes: Act*. Clemens Lode Verlag e.K. ISBN: 978-39-4558-623-5.

— (tba). *Philosophy for Heroes: Persona*. Clemens Lode Verlag e.K. ISBN: 978-39-4558-624-2.

Ramachandran, V. S. (2003). *The Emerging Mind*. Profile Books(GB); UK ed. edition. ISBN: 978-1861973030.

Ramachandran, V. S. and Sandra Blakeslee (1999). *Phantoms in the Brain: Probing the Mysteries of the Human Mind*. William Morrow Paperbacks. ISBN: 978-0688172176.

Rand, Ayn, Harry Binswanger, and Leonard Peikoff (1990). *Introduction to Objectivist Epistemology*. Expanded 2nd ed. New American Library, New York, N.Y. ISBN: 04-5201-030-6.

The Author

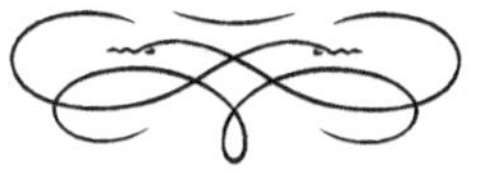

Clemens Lode studied computer science at the KIT Karlsruhe, and works as an author as well as an agile coach for teams throughout Europe. He lives in Düsseldorf (Germany). You can follow him on Facebook (https://fb.me/ClemensLode) or Twitter (https://www.twitter.com/ClemensLode), or just drop him a line (clemens@lode.de).

What I cannot create, I do not understand.

—Richard Feynman

Glossary

A

Allocortex • The *allocortex* is part of the cerebral cortex (the *neocortex* is the other part) and consists of the olfactory system and the hippocampus.

Amygdala • The *amygdala* is the brain's value and emotion center. It helps with evaluating thought patterns of the basal ganglia depending on the context instead of the mere strength of the signal. It also connects the brain with the hippothalamus, providing a bridge to the hormonal system.

Angular gyrus • The *angular gyrus* combines visual, auditory, and somatosensory information and puts things into relationship with each other. The *left angular gyrus* deals with relationships in the external world, especially relating to words and letters, and the *right angular gyrus* deals with the relationship between the self and the external world.

Attention • *Attention* is the brain's process of limiting alternative thought patterns, then increasing the most dominant thought pattern's strength. It is like a simple majority rule: the most successful thought pattern gets all the resources while other thought patterns are suppressed. While we can jump back and forth between different thoughts, we cannot have two dominant thought patterns at the same time.

Attention schema • The *attention schema* is a model the brain creates of the process of attention. It allows access to the working memory in order to be able to intervene before an action is taken.

Awareness • *Awareness* is a description of the process of attention. Something can grab the attention of your brain, but to talk about it, you need a model of what is happening in your brain. Awareness is such a model.

Awareness schema • The *awareness schema* is a model the brain creates of the process of awareness. With the awareness schema, the brain can write to the working memory to influence what the brain will focus on next. It also allows the brain to imagine alternative, past, or future scenarios.

B

Basal ganglia • The *basal ganglia* are a part of the brain that, like a referee, arbitrate decisions by the neural committees. Also, like an orchestra conductor, they coordinate the sequence of entire motor programs. In both cases, they do not make decisions but merely provide rules and structure.

Blindsight • Someone suffering from *blindsight* reports that he cannot see. However, experiments show that he can react to visual cues. As a result of damage in the visual cortex, information from the retina arrives in the midbrain but does not undergo conscious processing through the visual cortex.

Body schema • The *body schema* is the brain's simplified description of the status of the body. It is built by correlating what we see and feel our body is doing with signals that the brain sends to the muscles.

Broca's area • *Broca's area* is a brain part located in the left side of the frontal lobe and connected to Wernicke's area. It is responsible for the *production* of speech. Damage to Broca's area leads to a person unable to find the words to express what he wants to say. The homologous area in the right hemisphere deals with non-verbal communication.

Buddhism • *Buddhists* (Buddha lived 563–483 BC or 480–400 BC depending on the source) believe in a difference between brain and consciousness, with consciousness being compared to a light that shines on thoughts. The mind dies with the bodily death while "you" are reborn into a new being with no memories of your previous life.

C

Capgras delusion • A person suffering from *Capgras delusion* can recognize people who are close to him, but he thinks they have been replaced by clones or doppelgangers. One cause for this condition is a damaged or missing connection from the amygdala that leads to a person having no emotional connection to those whom he sees.

Cartesian theater • The *Cartesian theater* is a term coined by Daniel Dennett to criticize most of the contemporary explanations of consciousness. At their core, such explanations all share the view that there is some sort of miniature person ("homunculus") or entity within the brain looking at what we are looking at—an idea which ends in an infinite series of subsequently ever-smaller Cartesian theaters with ever-smaller homunculi.

Cerebellum • The *cerebellum* is the brain part that helps with coordination of complex behavior. It provides a set of motor programs the brain can choose from repeatedly for similar actions (even in time-critical situations). With the help of the cerebellum we can, for example, walk or bicycle without having to consciously think of each movement.

Cerebral cortex • The *cerebral cortex* is the outer layer of the cerebrum. It contains most of the neurons of the brain.

Cerebral hemispheres • The *cerebral hemispheres* consist of the *occipital lobe*, the *temporal lobe*, the *parietal lobe*, and the *frontal lobe*. The two hemispheres are joined by the *corpus callosum*.

Cerebrum • The *cerebrum* includes the neocortex (the cerebral hemispheres), and the allocortex (the hippocampus, the basal ganglia, and the olfactory bulb).

Consciousness • *Consciousness* is an umbrella term for the brain's abilities to process sense data, focus on something, be aware of something, have the ability to process high-level information (attention schema), be able to control awareness (awareness schema), and reflect on abstract information (philosophy and science).

Copenhagen interpretation • In the *Copenhagen interpretation* of quantum mechanics, an observer is required for the wave function to collapse. Without observation, the wave function never collapses and never becomes a particle. While the interpretation does not mention consciousness as such (measurements by a device are observations, too), it raises the question of who observes the observer, resulting in an infinite loop.

Corpus callosum • The *corpus callosum* connects the left and right brain hemispheres, coordinating tasks requiring both sides.

D

Declarative memory • *Declarative memory* connects one memory with another. For example, the amygdala maps thalamic sense input to emotions; the hippocampus maps places with each other for orientation; and the neocortex maps concepts with other concepts (a tree is a plant).

E

Emergent property • An *emergent property* is a property of a system that emerges only when its parts are combined or interact with each other. Individual parts of that system do not have the emergent property themselves. For example, the division of labor in ant colonies allows the ants to be more efficient than if individual ants fended for themselves.

Encephalization quotient • The *encephalization quotient* (EQ) is a measure of relative brain size and is often used to convey how small or large a species' brain is compared to that of other species of similar body size.

Episodic buffer • The *episodic buffer* is responsible for remembering the sequence of events, including the last state of an entity.

Evil demon • Descartes' *evil demon* is a thought experiment to differentiate the immaterial mind from the brain. His "evil demon" is an entity that could make any changes to the material world without anyone being aware of such changes. Descartes' assumption was that by examining the remaining things we could rely on, we would discover what the immaterial mind is.

Executive functions • The *executive functions* represent a series of means by which the prefrontal cortex can suppress thought patterns in other parts of the brain. While winners in the neural competition are selected with the help of the basal ganglia, the prefrontal cortex can counteract those decisions in favor of other actions. The prefrontal cortex makes these decisions based on its models. For example, stealing goes against the norms of society, so the prefrontal cortex suppresses the (utilization behavior of the parietal lobe's) urge to grab someone else's property.

F

Frontal lobe • The *frontal lobe* of the neocortex deals with running a simplified simulation of the world. It provides us the ability to plan and evaluate actions and their future impact. The somatosensory area of the parietal lobe is directly adjacent to the frontal lobe.

G

Gyrus • A *gyrus* is a fold or ridge in the cerebral cortex.

H

Hard problem of consciousness • The *hard problem of consciousness* asks the question where the subjective experience of consciousness comes from. It is "hard" as there are no known ways of detecting this experience objectively without relying on the subjective claims of an individual (or ourselves).

Hemispatial neglect • Someone suffering from *hemispatial neglect* lacks consciousness of half of his visual field. The person is not aware that his vision is impaired in any way, making the condition different from blindness in one eye. People with this condition have to learn abstract strategies as a way of coping.

Hippocampus • The brain's *hippocampus* provides us with a mental map for navigation. It also builds temporal relationships between places, allowing us to determine, for example, which areas in our environment we have already foraged and in which areas the plants have regrown. The *hippocampus* and the *olfactory system* (sense of smell) make up the *allocortex*.

Homunculus argument fallacy • The *homunculus argument* is the fallacy of trying to explain consciousness by another (smaller) conscious person (the "homunculus") observing and steering you. The problem with this explanation is that it remains unexplained how this smaller homunculus subsequently experiences consciousness.

I

Indian idealism (Vedanta) • In the *Indian idealist* worldview, there is but a single consciousness and our experience of the world as separate beings or consciousnesses is an illusion.

Intuition • Your initial evaluation of a situation is called *intuition*. It is the first thing that comes to your mind without going through conscious deliberation or reasoning.

L

Lateral geniculate nucleus • The *lateral geniculate nucleus* (LGN) is part of the thalamus and relays information from the retinas (via the optic chiasm) to the visual cortex. It pre-processes some of the information, for example, combining red, green, and blue photoreceptor cells into colors.

Lobe • A *lobe* is an anatomical division or extension of an organ.

Loop of consciousness • The *loop of consciousness* refers to a network of connections in the brain from the thalamus to the cortex, then to the basal ganglia, and then back to the thalamus (also called the *cortico-basal ganglia-thalamo-cortical loop*). It is believed that this is the source of our subjective experience of the world.

M

Many-minds interpretation • The *many-minds interpretation* of quantum mechanics is similar to the many-worlds interpretation, in which the universe splits into infinite universes. In the many-minds interpretation, the split of the universe happens with each thought for each individual brain, instead of with each measurement (as in the many-worlds interpretation). Consequently, one's consciousness splits into many consciousnesses whenever a decision was made.

Materialism • In the *materialist* worldview, there is no separate "mind." Instead, everything can be explained by a single substance (matter).

Mechanistic theory • A *mechanistic theory* is a theory that explains a system as if it were a computer program or a mechanical machine with gearwheels: you can understand it by tracing the input to the output. Pedagogically, a mechanistic theory is of great value as we can explain the essential workings of a system step by step using a diagram with boxes and arrows.

Mind-brain dualism • The philosophical view of *mind-brain dualism* (or also *mind-body dualism* with "body" including our brain) states that what we call the (immaterial) mind is separate from the material world (the body, including the brain).

Mirror test • The *mirror test* evaluates the ability of an animal to recognize itself in a mirror after a researcher has secretly added a blot of coloring to the animal's body and put the animal in front of a mirror. If the animal starts investigating the blot of color on its own body instead of on the mirror image, it passes the test (because that indicates the animal realizes it is the same creature that it sees in the mirror).

Model • The *model* of an entity is a simplified simulation of that entity. It consists of the entity's concepts and its properties, as well as some of the entity's measurements.

Monism • The philosophical view of *Monism* is that everything that exists (including what we call mind or consciousness) can be traced back to a single fabric of the universe. The consequence of this view is that the mind must be a result of a (mechanistic materialist) process and that mind and matter are not two separate things.

Monothematic delusion • A *monothematic delusion* is a delusion focused on a single topic. A delusion is a firm belief that cannot be swayed by rational arguments. It is distinct from false beliefs that are based on false or incomplete information, erroneous logical conclusions, or perceptual problems.

N

Neocortex • The *neocortex* is the newest part of the mammalian brain and consists of the *cerebral hemispheres*. Its main tasks are focus, language, long-term planning, and modelling of the world. It can generate strategies that involve detours if goal-directed behavior is not successful (for example, going around a fence instead of trying to get through it).

Nerve net • A *nerve net* is nervous system without a central organization. This means that any signal (that exceeds a certain signal strength) the nerve net receives through its senses causes a singular reaction. Hydras have this kind of nervous system and use it to contract their body when prey touches their tentacles.

Neutral monism • In the *neutral monist* worldview, while both mind and matter exist, they both stem from a third, undefined substance.

Nocturnal bottleneck hypothesis • The *nocturnal bottleneck hypothesis* posits that many mammalian traits were adaptions to moving into a niche to become nocturnal animals and evade the dominant dinosaurs.

O

Object permanence • The ability of *object permanence* allows us to track predicted positions or movements of objects even after they have vanished from our field of view. By running a simplified simulation of the world, we are aware that a tiger that has jumped behind a tree is still there.

Objectivism • *Objectivism* (founded by Ayn Rand, 1902–1982) is a monist philosophy that recognizes a distinction between the brain and a "prime mover" that has a power of whether to focus the brain or not.

Occipital lobe • The *occipital lobe* is part of the *neocortex* and contains the *visual cortex* which is responsible for processing visual sense data.

P

Parietal lobe • The *parietal lobe* is the part of the neocortex that deals with the "where," especially the location of entities, the "how," as well as touch perception.

Philosophical zombie • A *philosophical zombie* is a hypothetical being that acts exactly like a human but lacks the inner conscious experience. The existence of a philosophical zombie is used as an argument against consciousness being a mechanistic process: if we are but a mechanistic machine, why would we need a subjective, conscious experience? This argument is addressed by pointing out the necessary evolutionary advantage of having a conscious experience, which is the ability to focus and to explain oneself to others.

Phonological loop • The *phonological loop* is an acoustic memory system of the brain that holds spoken words or sounds. It involves (among other brain parts, see Buchsbaum and D'Esposito, 2008) Broca's area and Wernicke's area.

Prefrontal cortex • The *prefrontal cortex* is part of the frontal lobe and can be understood as running a simulation of the world. It monitors social relationships, keeps track of objects when they are no longer visible (object permanence), and helps with the pursuit of long-term goals. It has only indirect connections to brain parts dealing with actions or sense perception.

Premotor cortex • The *premotor cortex* is part of the frontal lobe and prepares motor programs to be executed by the adjacent *primary motor cortex*. It has strong connections to the *superior parietal lobe* which provides a model of the current state of the limbs.

Primary motor cortex • The *primary motor cortex* is part of the frontal lobe and is directly adjacent to the primary somatosensory cortex of the parietal cortex. This way, it can directly process data from our sense of touch to better control movements. The *primary motor cortex* connects to the *brainstem* and *spinal cord* (via the *upper motor neurons*) which in turn connect to the muscles (via the lower motor neurons).

Primary somatosensory cortex • The *primary somatosensory cortex* in the parietal lobe deals with the processing of tactile sense input. Each body part is represented in the primary somatosensory cortex. The size of each representation correlates with the sensitivity of the body part to tactile stimulation (for example, lips and hands have a larger representation than other body parts).

Q

Quantum mind • The term *quantum mind* refers to a collection of theories that consider quantum mechanics as the basis of consciousness.

R

Rubber hand illusion • The *rubber hand illusion* can be evoked by brushing both a fake rubber hand and the real hand of a human participant. If only the fake rubber hand is visible, the brain will assume it is the real hand and incorporate it into its body schema.

S

Savannah hypothesis • The *savannah hypothesis* states that early humans evolved on the savannah and that many of the modern human's traits are a result of this adaption.

Sense data • *Sense data* are informations, converted to a form usable by cognition, about an effect registered by a sensory organ.

Split-brain syndrome • The *split-brain syndrome* can occur when the *corpus callosum* is damaged. This leads to problems with communication between the left and right brain hemispheres and complicates some tasks requiring both sides.

Subjective idealism • In the *subjective idealist* worldview, there is no such thing as "matter." Instead, everything is but perception, mind, or "consciousness," and nothing exists but human minds and gods.

Sulcus • A *sulcus* is a groove in the cerebral cortex.

Superior colliculus • The *superior colliculus* or *optic tectum* (in non-mammals) helps the eyes to track objects, and controls blinking, pupillary, and head-turning reflexes.

Superior parietal lobe • The *superior parietal lobe* deals with creating and maintaining a mental representation of the internal state of the body.

Supervised learning • Using *supervised learning*, a brain (or computer) can improve its response to a situation with each new encounter. For example, a dog can learn to sit or roll over on command by getting positive rewards for doing so during training.

Supramarginal gyrus • The *supramarginal gyrus* (SMG) creates an internal representation of the *limbs* of the body, similar to the adjacent superior parietal lobe (which creates a representation of the *internal state* of the body). It supports tool use (left SMG) and interpreting the emotional state of other people based on their postures and gestures (right SMG).

T

Tabula rasa • *Tabula rasa*, (meaning "blank slate"), refers to the view that we are born without any innate knowledge and that our minds can create knowledge only with the help of sense data.

Temporal lobe • The *temporal lobe* is the part of the neocortex that deals with the "what": long-term memory, and object, face, and speech recognition.

Thalamus • The *thalamus* integrates different sensory information and relays the information to other brain parts. For example, it combines sense data from the retinas' cones into colors, or calculates three-dimensional information from the two-dimensional images from both eyes.

Theory of mind • The concept of *theory of mind* refers to the ability to imagine what other people (or animals) are thinking and what they know. This provides a significant advantage for hunting (predicting whether the prey can see or otherwise sense you), or social interaction (teaching, trading, and even lying, etc.).

Temporoparietal junction • The *temporoparietal junction* is a brain area between the temporal lobe and the parietal lobe, integrating data from the thalamus, as well as the limbic system, and the visual, auditory, and somatosensory systems. Combining the "what" and the "where" information, it creates a distinction between "self" and "other."

Turing test • In 1950, Alan Turing proposed the *Turing test* to assess whether or not a machine is intelligent. In the test, a human participant would observe a text chat between a computer and a human. The machine would pass if the observer could not tell who was the machine and who was the human.

U

Uncanny valley effect • The *uncanny valley effect* refers to the negative reaction to dolls or robots that are very (but not fully) human-like. Possible reasons for this reaction are an inbuilt instinct to avoid corpses, and the inner conflict and discomfort of switching back and forth between seeing a being as fully human or a lifeless object.

Unsupervised learning • Using *unsupervised learning*, a brain (or computer) can build a concept by analyzing several sense perceptions, finding commonalities, and dropping measurements. For example, unsupervised learning could be used to form the concept "table" by encountering several different tables and finding out that they share properties like having a table-top, the form, material, and size of the table-top, and the number of table legs.

Viso-spatial scratchpad • The *visuo-spatial scratchpad* is responsible for temporarily storing visual and spatial information (involving the occipital lobes).

Von Neumann-Wigner interpretation • The *von Neumann-Wigner interpretation* of quantum mechanics tries to solve the loop of the Copenhagen interpretation by stating that consciousness is outside of the quantum world and is the final observer that collapses the wave function. For consciousness itself to exist, no observation of consciousness would be needed.

W

Wernicke's area • *Wernicke's area* is located in the left side of the temporal lobe. Its function is the *comprehension* of speech. Damage to Wernicke's area leads to people losing the ability to form meaningful sentences. It is connected to Broca's area, which is responsible for muscle activation to *produce* speech.

Working memory • *Working memory* is the collection of a number of different (limited) short-term memory systems in the brain. The prefrontal cortex has access to the working memory.

Bibliography

Andres, Michael et al. (2017). "The left supramarginal gyrus contributes to finger positioning for object use: a neuronavigated transcranial magnetic stimulation study." In: *European Journal of Neuroscience* 46.12, pp. 2835–2843. DOI: 10.1111/ejn.13763. URL: https://onlinelibrary.wiley.com/doi/abs/10.1111/ejn.13763.

Apergis-Schoute, Annemieke M. et al. (2017). "Neural basis of impaired safety signaling in Obsessive Compulsive Disorder." In: *Proceedings of the National Academy of Sciences.* ISSN: 0027-8424. DOI: 10.1073/pnas.1609194114. URL: https://www.pnas.org/content/early/2017/02/28/1609194114.

Arain, Mariam et al. (2013). "Maturation of the adolescent brain." eng. In: *Neuropsychiatric disease and treatment* 9, pp. 449–461. ISSN: 1176-6328. DOI: 10.2147/NDT.S39776. URL: https://pubmed.ncbi.nlm.nih.gov/23579318.

Ari, Csilla and Dominic D'Agostino (2016). "Contingency checking and self-directed behaviors in giant manta rays: Do elasmobranchs have self-awareness?" In: *Journal of Ethology* 34. DOI: 10.1007/s10164-016-0462-z.

Arzi, Anat et al. (2020). "Olfactory sniffing signals consciousness in unresponsive patients with brain injuries." In: *Nature.* ISSN: 1476-4687. DOI: 10.1038/s41586-020-2245-5. URL: https://doi.org/10.1038/s41586-020-2245-5.

Arzy, Shahar et al. (2006). "Neural basis of embodiment: distinct contributions of temporoparietal junction and extrastriate body area." In: *The Journal of neuroscience* 26.31, pp. 8074–8081. DOI: 10.1523/JNEUROSCI.0745-06.2006. URL: http://infoscience.epfl.ch/record/154875.

Baird, A. et al. (2002). "Frontal lobe activation during object permanence: Data from near-infrared spectroscopy." In: *NeuroImage* 16, pp. 1120–1125. DOI: 10.1006/nimg.2002.1170.

Baumgartner, Thomas et al. (2015). "Neuroanatomy of intergroup bias: A white matter microstructure study of individual differences." In: *NeuroImage* 122, pp. 345–354. ISSN: 1053-8119. DOI: https://doi.org/10.1016/j.neuroimage.2015.08.011. URL: http://www.sciencedirect.com/science/article/pii/S105381191500720X.

Beagle, Peter S. (1991). *The Last Unicorn.* Roc Trade. ISBN: 978-0451450524.

Blanke, Olaf et al. (2002). "Stimulating Illusory Own-Body Perceptions." In: *Nature* 419, pp. 269–270. DOI: 10.1038/419269a.

Bostan, Andreea C. and Peter L. Strick (2018). "The basal ganglia and the cerebellum: nodes in an integrated network." In: *Nature Reviews Neuroscience* 19.6, pp. 338–350. ISSN: 1471-0048. DOI: 10.1038/s41583-018-0002-7. URL: https://doi.org/10.1038/s41583-018-0002-7.

Botvinick, Matthew and Jonathan Cohen (1998). "Rubber hands 'feel' touch that eyes see." In: *Nature* 391. ISSN: 1476-4687. DOI: 10.1038/35784. URL: https://doi.org/10.1038/35784.

Buchsbaum, Bradley R. and Mark D'Esposito (2008). "The Search for the Phonological Store: From Loop to Convolution." In: *Journal of Cognitive Neuroscience* 20.5, pp. 762–778. DOI: 10.1162/jocn.2008.20501. URL: https://doi.org/10.1162/jocn.2008.20501.

Calvin, William H. (1996). *The Cerebral Code: Thinking a Thought in the Mosaics of the Mind.* A Bradford Book. ISBN: 978-0262531542.

— (2007). *A Brief History of the Mind: From Apes to Intellect and Beyond.* Oxford University Press. ISBN: 978-01-9518-248-4.

Carius, Daniel et al. (2020). "Characterizing cortical hemodynamic changes during climbing and its relation to climbing expertise." In: *Neuroscience Letters* 715, p. 134604. ISSN: 0304-3940. DOI: https://doi.org/10.1016/j.neulet.2019.134604. URL: http://www.sciencedirect.com/science/article/pii/S0304394019307074.

Carta, Ilaria et al. (2019). "Cerebellar modulation of the reward circuitry and social behavior." In: *Science* 363.6424. ISSN: 0036-8075. DOI: 10.1126/science.aav0581. URL: https://science.sciencemag.org/content/363/6424/eaav0581.

Chai, Wen Jia, Aini Ismafairus Abd Hamid, and Jafri Malin Abdullah (2018). "Working Memory From the Psychological and Neurosciences Perspectives: A Review." eng. In: *Frontiers in psychology* 9, p. 401. ISSN: 1664-1078. DOI: 10.3389/fpsyg.2018.00401. URL: https://pubmed.ncbi.nlm.nih.gov/29636715.

Changizi, Mark A. et al. (2008). "Perceiving the Present and a Systematization of Illusions." In: *Cognitive Science* 32.3, pp. 459–503. DOI: 10.1080/03640210802035191. URL: https://onlinelibrary.wiley.com/doi/abs/10.1080/03640210802035191.

Chen, Lu et al. (2020). "Identifying and Interpreting Apparent Neanderthal Ancestry in African Individuals." In: *Cell* 180.4, pp. 677–687. ISSN: 0092-8674. DOI: https://doi.org/10.1016/j.cell.2020.01.012. URL: http://www.sciencedirect.com/science/article/pii/S0092867420300593.

Chen, Zhanqi et al. (2018). "Prolonged milk provisioning in a jumping spider." In: *Science* 362.6418, pp. 1052–1055. ISSN: 0036-8075. DOI: 10.1126/science.aat3692. URL: https://science.sciencemag.org/content/362/6418/1052.

Cleeremans, Axel (2011). "The Radical Plasticity Thesis: How the Brain Learns to be Conscious." In: *Frontiers in Psychology* 2, p. 86. ISSN: 1664-1078. DOI: 10.3389/fpsyg.2011.00086. URL: https://www.frontiersin.org/article/10.3389/fpsyg.2011.00086.

Conant, Roger C. and W. Ross Ashby (1970). "Every good regulator of a system must be a model of that system." In: *International Journal of Systems Science* 1.2, pp. 89–97. DOI: 10.1080/00207727008920220. URL: https://doi.org/10.1080/00207727008920220.

Danjo, Teruko, Taro Toyoizumi, and Shigeyoshi Fujisawa (2018). "Spatial representations of self and other in the hippocampus." In: *Science* 359.6372, pp. 213–218. ISSN: 0036-8075. DOI: 10.1126/science.aao3898. URL: https://science.sciencemag.org/content/359/6372/213.

Dawkins, Richard (1990). *The Selfish Gene*. Oxford University Press. ISBN: 978-01-9286-092-7.

— (1997). *Climbing Mount Improbable*. W. W. Norton and Company. ISBN: 978-0393316827.

Denion, Eric et al. (2014). "Eye motion increases temporal visual field extent." In: *Acta Ophthalmologica* 92.3, e200–e206. DOI: 10.1111/aos.12106. URL: https://onlinelibrary.wiley.com/doi/abs/10.1111/aos.12106.

Dennett, Daniel (1993). *Consciousness Explained*. Penguin. ISBN: 978-0140128673.

Deschamps, Isabelle, Shari R. Baum, and Vincent L. Gracco (2014). "On the role of the supramarginal gyrus in phonological processing and verbal working memory: Evidence from rTMS studies." In: *Neuropsychologia* 53, pp. 39–46. ISSN: 0028-3932. DOI: https://doi.org/10.1016/j.neuropsychologia.2013.10.015. URL: http://www.sciencedirect.com/science/article/pii/S0028393213003655.

Domínguez-Rodrigo, M. (2014). "Is the Savanna Hypothesis a Dead Concept for Explaining the Emergence of the Earliest Hominins?" In: *Current Anthropology* 55.1, pp. 59–81. DOI: 10.1086/674530. URL: https://doi.org/10.1086/674530.

Donahue, Chad J. et al. (2018). "Quantitative assessment of prefrontal cortex in humans relative to nonhuman primates." In: *Proceedings of the National Academy of Sciences* 115.22, E5183–E5192. ISSN: 0027-8424. DOI: 10.1073/pnas.1721653115. URL: https://www.pnas.org/content/115/22/E5183.

Dutschke, Lars Levi et al. (2017). "Brain Tumor-Associated Psychosis and Spirituality—A Case Report." In: *Frontiers in Psychiatry* 8, p. 237. ISSN: 1664-0640. DOI: 10.3389/fpsyt.2017.00237. URL: https://www.frontiersin.org/article/10.3389/fpsyt.2017.00237.

Ehrsson, H. Henrik (2007). "The Experimental Induction of Out-of-Body Experiences." In: *Science* 317.5841, p. 1048. ISSN: 0036-8075. DOI: 10.1126/science.1142175. URL: https://science.sciencemag.org/content/317/5841/1048.

Ejsmond, Maciej Jan, Jacek Radwan, and Anthony B. Wilson (2014). "Sexual selection and the evolutionary dynamics of the major histocompatibility complex." In: *Proceedings of the Royal Society B: Biological Sciences* 281.1796. DOI: 10.1098/rspb.2014.1662. URL: https://royalsocietypublishing.org/doi/abs/10.1098/rspb.2014.1662.

Elsaesser, Rebecca and Jacques Paysan (2007). "The sense of smell, its signalling pathways, and the dichotomy of cilia and microvilli in olfactory sensory cells." In: *BMC neuroscience* 8 Suppl 3, S1. DOI: 10.1186/1471-2202-8-S3-S1.

Feuillet, Lionel, Henry Dufour, and Jean Pelletier (2007). "Brain of a white-collar worker." In: *Lancet* 370, p. 262. DOI: 10.1016/S0140-6736(07)61127-1.

Feynman, Richard P. (2012). *Character of Physical Law*. Penguin. ISBN: 978-01-4017-505-9.

Fjell, Anders M. et al. (2013). "High-Expanding Cortical Regions in Human Development and Evolution Are Related to Higher Intellectual Abilities." In: *Cerebral Cortex* 25.1, pp. 26–34. ISSN: 1047-3211. DOI: 10.1093/cercor/bht201. URL: https://doi.org/10.1093/cercor/bht201.

Fulford, Jon et al. (2018). "The neural correlates of visual imagery vividness - An fMRI study and literature review." In: *Cortex* 105. The Eye's Mind - visual imagination, neuroscience and the humanities, pp. 26–40. ISSN: 0010-9452. DOI: https://doi.org/10.1016/j.cortex.2017.09.014. URL: http://www.sciencedirect.com/science/article/pii/S0010945217303209.

Furness, J. B., J. J. Cottrell, and D. M. Bravo (2015). "Comparative physiology of digestion." In: *Journal of Animal Science* 93.2, pp. 485–491. ISSN: 0021-8812. DOI: 10.2527/jas.2014-8481. URL: https://doi.org/10.2527/jas.2014-8481.

Gallagher, H.L et al. (2000). "Reading the mind in cartoons and stories: an fMRI study of theory of mind in verbal and nonverbal tasks." In: *Neuropsychologia* 38.1, pp. 11–21. ISSN: 0028-3932. DOI: https://doi.org/10.1016/S0028-3932(99)00053-6. URL: http://www.sciencedirect.com/science/article/pii/S0028393299000536.

Gallup, Gordon G. (1970). "Chimpanzees: Self-Recognition." In: *Science* 167.3914, pp. 86–87. ISSN: 0036-8075. DOI: 10.1126/science.167.3914.86. URL: https://science.sciencemag.org/content/167/3914/86.

Gorno-Tempini, M L et al. (1998). "The neural systems sustaining face and proper-name processing." In: *Brain* 121.11, pp. 2103–2118. ISSN: 0006-8950. DOI: 10.1093/brain/121.11.2103. URL: https://doi.org/10.1093/brain/121.11.2103.

Graziano, Michael (2015a). *Consciousness and the Social Brain*. eng. First Edition. New York: Oxford University Press. ISBN: 978-0190263195.

— (2015b). *Michael Graziano, Closer to Truth*. [online; last accessed March 7th, 2020]. URL: https://www.youtube.com/watch?v=gsRb5PJcBP4.

— (2017). "The Attention Schema Theory: A Foundation for Engineering Artificial Consciousness." In: *Frontiers in Robotics and AI* 4, p. 60. ISSN: 2296-9144. DOI: 10.3389/frobt.2017.00060. URL: https://www.frontiersin.org/article/10.3389/frobt.2017.00060.

— (2019). *Rethinking Consciousness - A Scientific Theory of Subjective Experience*. English. W. W. Norton & Company. ISBN: 978-0393652611.

Graziano, Michael and Taylor W Webb (2015). "The attention schema theory: A mechanistic account of subjective awareness." In: *Frontiers in psychology* 6, p. 500. DOI: 10.3389/fpsyg.2016.00500.

Guardia, Dewi et al. (2012). "Imagining One's Own and Someone Else's Body Actions: Dissociation in Anorexia Nervosa." In: *PLOS ONE* 7.8, pp. 1–9. DOI: 10.1371/journal.pone.0043241. URL: https://doi.org/10.1371/journal.pone.0043241.

Harpaz, Yuval, Yechiel Levkovitz, and Michal Lavidor (2009). "Lexical ambiguity resolution in Wernicke's area and its right homologue." In: *Cortex* 45.9, pp. 1097–1103. ISSN: 0010-9452. DOI: https://doi.org/10.1016/j.cortex.2009.01.002. URL: http://www.sciencedirect.com/science/article/pii/S001094520900029X.

Hartwigsen, Gesa et al. (2010). "Phonological decisions require both the left and right supramarginal gyri." In: *Proceedings of the National Academy of Sciences* 107.38, pp. 16494–16499. ISSN: 0027-8424. DOI: 10.1073/pnas.1008121107. URL: https://www.pnas.org/content/107/38/16494.

Hirosawa, Rikuei et al. (2013). "Reduced dorsolateral prefrontal cortical hemodynamic response in adult obsessive-compulsive disorder as measured by near-infrared spectroscopy during the verbal fluency task." In: *Neuropsychiatric disease and treatment* 9. 23874098[pmid], pp. 955–962. ISSN: 1176-6328. DOI: 10.2147/NDT.S45402. URL: https://pubmed.ncbi.nlm.nih.gov/23874098.

Horowitz, Alexandra (2017). "Smelling themselves: Dogs investigate their own odours longer when modified in an olfactory mirror test." In: *Behavioural Processes* 143, pp. 17–24. ISSN: 0376-6357. DOI: https://doi.org/10.1016/j.beproc.2017.08.001. URL: http://www.sciencedirect.com/science/article/pii/S0376635717300104.

Imura, Tomoko and Masaki Tomonaga (2013). "Differences between chimpanzees and humans in visual temporal integration." In: *Scientific reports* 3, p. 3256. ISSN: 2045-2322. DOI: 10.1038/srep03256. URL: https://pubmed.ncbi.nlm.nih.gov/24247153.

Inoue, Sana and Tetsuro Matsuzawa (2007). "Working memory of numerals in chimpanzees." In: *Current Biology* 17.23, R1004–R1005. URL: http://dx.doi.org/10.1016/j.cub.2007.10.027.

Johns, Paul (2014). "Chapter 3 - Functional neuroanatomy." In: *Clinical Neuroscience.* Ed. by Paul Johns. Churchill Livingstone, pp. 27–47. ISBN: 978-0-443-10321-6. DOI: https://doi.org/10.1016/B978-0-443-10321-6.00003-5. URL: http://www.sciencedirect.com/science/article/pii/B9780443103216000035.

Kaas, Jon H., ed. (2017). *Evolution of Nervous Systems.* 2nd ed. Vol. 1-4. editor-in-chief, Jon H. Kaas.ill. (some col.), ports. ; 27 cm.vol. 1. vol. 2. vol. 3. vol. 4. Amsterdam; Boston: Academic Press, p. 2007. ISBN: 9780123925640. URL: https://www.elsevier.com/books/evolution-of-nervous-systems/kaas/978-0-12-804042-3.

Kaminski, Gwenaël et al. (2009). "Human ability to detect kinship in strangers' faces: effects of the degree of relatedness." In: *Proceedings of the Royal Society B: Biological Sciences* 276.1670, pp. 3193–3200. DOI: 10.1098/rspb.2009.0677. URL: https://royalsocietypublishing.org/doi/abs/10.1098/rspb.2009.0677.

Keller, Helen (2010). *The World I Live In and Optimism: A Collection of Essays.* Dover Publications. ISBN: 978-0486473673.

Kerney, Max et al. (2017). "The coevolution of play and the cortico-cerebellar system in primates." In: *Primates* 58.4, pp. 485–491. ISSN: 1610-7365. DOI: 10.1007/s10329-017-0615-x. URL: https://doi.org/10.1007/s10329-017-0615-x.

Krall, S et al. (2014). "The role of the right temporoparietal junction in attention and social interaction as revealed by ALE meta-analysis." In: *Brain structure & function* 220. DOI: 10.1007/s00429-014-0803-z.

Kuhlwilm, Martin et al. (2016). "Ancient gene flow from early modern humans into Eastern Neanderthals." In: *Nature* 530. DOI: 10.1038/nature16544.

Lenggenhager, Bigna et al. (2007). "Video Ergo Sum: Manipulating Bodily Self-Consciousness." In: *Science* 317.5841, pp. 1096–1099. ISSN: 0036-8075. DOI: 10.1126/science.1143439. URL: https://science.sciencemag.org/content/317/5841/1096.

Lode, Clemens (2016). *Philosophy for Heroes: Knowledge.* Clemens Lode Verlag e.K. ISBN: 978-39-4558-621-1.

—   (2018a). *Philosophy for Heroes: Continuum.* Clemens Lode Verlag e.K. ISBN: 978-39-4558-622-8.

—   (2018b). *Writing Better Books the Agile Way.* Clemens Lode Verlag e.K. ISBN: 978-3-945586-83-9.

— (2019). *Better Books with LaTeX the Agile Way*. Clemens Lode Verlag e.K. ISBN: 978-3-945586-48-8.

— (2020). *Philosophy for Heroes: Act*. Clemens Lode Verlag e.K. ISBN: 978-39-4558-623-5.

— (tba). *Philosophy for Heroes: Persona*. Clemens Lode Verlag e.K. ISBN: 978-39-4558-624-2.

Luncz, Lydia V. et al. (2017). "Technological Response of Wild Macaques (Macaca fascicularis) to Anthropogenic Change." In: *International Journal of Primatology* 38.5, pp. 872–880. ISSN: 1573-8604. DOI: 10.1007/s10764-017-9985-6. URL: https://doi.org/10.1007/s10764-017-9985-6.

Lussanet, Marc de and Jan Osse (2012). "An ancestral axial twist explains the contralateral forebrain and the optic chiasm in vertebrates." In: *Animal Biol.* 62, pp. 193–216. DOI: 10.1163/157075611X617102.

MacIver, K. et al. (2008). "Phantom limb pain, cortical reorganization and the therapeutic effect of mental imagery." In: *Brain* 131.8, pp. 2181–2191. ISSN: 0006-8950. DOI: 10.1093/brain/awn124. URL: https://doi.org/10.1093/brain/awn124.

Marasco, Paul D. et al. (2018). "Illusory movement perception improves motor control for prosthetic hands." In: *Science Translational Medicine* 10.432. ISSN: 1946-6234. DOI: 10.1126/scitranslmed.aao6990. URL: https://stm.sciencemag.org/content/10/432/eaao6990.

Marshall, John C. and Peter W. Halligan (1988). "Blindsight and insight in visuo-spatial neglect." In: *Nature* 336. ISSN: 1476-4687. DOI: 10.1038/336766a0. URL: https://doi.org/10.1038/336766a0.

McAlonan, Kerry, James Cavanaugh, and Robert H. Wurtz (2006). "Attentional Modulation of Thalamic Reticular Neurons." In: *Journal of Neuroscience* 26.16, pp. 4444–4450. ISSN: 0270-6474. DOI: 10.1523/JNEUROSCI.5602-05.2006. URL: https://www.jneurosci.org/content/26/16/4444.

McCown, William (1989). "The Relationship Between Impulsivity, Empathy and Involvement in Twelve Step Self-help Substance Abuse Treatment Groups." In: *British Journal of Addiction* 84.4, pp. 391–393. DOI: 10.1111/j.1360-0443.1989.tb00582.x. URL: https://onlinelibrary.wiley.com/doi/abs/10.1111/j.1360-0443.1989.tb00582.x.

McKernan, M. G. and P. Shinnick-Gallagher (1997). "Fear conditioning induces a lasting potentiation of synaptic currents in vitro." In: *Nature* 390.6660, pp. 607–611. ISSN: 1476-4687. DOI: 10.1038/37605. URL: https://doi.org/10.1038/37605.

Miller, Greg (2007). "Out-of-Body Experiences Enter the Laboratory." In: *Science* 317.5841, pp. 1020–1021. ISSN: 0036-8075. DOI: 10.1126/science.317.5841.1020a. URL: https://science.sciencemag.org/content/317/5841/1020.1.

Minxha, Juri et al. (2020). "Flexible recruitment of memory-based choice representations by the human medial frontal cortex." In: *Science* 368.6498. ISSN: 0036-8075. DOI: 10.1126/science.aba3313. URL: https://science.sciencemag.org/content/368/6498/eaba3313.

Mishkin, Mortimer and Leslie G. Ungerleider (1982). "Contribution of striate inputs to the visuospatial functions of parieto-preoccipital cortex in monkeys." In: *Behavioural Brain Research* 6.1, pp. 57–77. ISSN: 0166-4328. DOI: https://doi.org/10.1016/0166-4328(82)90081-X. URL: http://www.sciencedirect.com/science/article/pii/016643288290081X.

Misselhorn, Jonas, Uwe Friese, and Andreas K. Engel (2019). "Frontal and parietal alpha oscillations reflect attentional modulation of cross-modal matching." In: *Scientific Reports* 9.1, p. 5030. ISSN: 2045-2322. DOI: 10.1038/s41598-019-41636-w. URL: https://doi.org/10.1038/s41598-019-41636-w.

Moosa, Mahdi Muhammad and S. M. Minhaz Ud-Dean (2010). "Danger Avoidance: An Evolutionary Explanation of Uncanny Valley." In: *Biological Theory* 5.1, pp. 12–14. ISSN: 1555-5550. DOI: 10.1162/BIOT_a_00016. URL: https://doi.org/10.1162/BIOT_a_00016.

Morgan, Andrew T., Lucy S. Petro, and Lars Muckli (2019). "Scene representations conveyed by cortical feedback to early visual cortex can be described by line drawings." In: *Jour-

nal of Neuroscience. ISSN: 0270-6474. DOI: 10.1523/JNEUROSCI.0852-19.2019. URL: https://www.jneurosci.org/content/early/2019/11/01/JNEUROSCI.0852-19.2019.

Morishima, Yosuke et al. (2012). "Linking Brain Structure and Activation in Temporoparietal Junction to Explain the Neurobiology of Human Altruism." In: *Neuron* 75.1, pp. 73–79. ISSN: 0896-6273. DOI: 10.1016/j.neuron.2012.05.021. URL: https://doi.org/10.1016/j.neuron.2012.05.021.

Mujica-Parodi, Lilianne R. et al. (2009). "Chemosensory Cues to Conspecific Emotional Stress Activate Amygdala in Humans." In: *PLOS ONE* 4.7, pp. 1–14. DOI: 10.1371/journal.pone.0006415. URL: https://doi.org/10.1371/journal.pone.0006415.

Murray, Elisabeth A., Steven P. Wise, and Kim S. Graham (2018). "Representational specializations of the hippocampus in phylogenetic perspective." In: *Neuroscience Letters* 680. New Perspectives on the Hippocampus and Memory, pp. 4–12. ISSN: 0304-3940. DOI: https://doi.org/10.1016/j.neulet.2017.04.065. URL: http://www.sciencedirect.com/science/article/pii/S0304394017303786.

Nakamura-Palacios, Ester M. et al. (2014). "Gray Matter Volume in Left Rostral Middle Frontal and Left Cerebellar Cortices Predicts Frontal Executive Performance in Alcoholic Subjects." In: *Alcoholism: Clinical and Experimental Research* 38.4, pp. 1126–1133. DOI: 10.1111/acer.12308. URL: https://onlinelibrary.wiley.com/doi/abs/10.1111/acer.12308.

Nattrass, Stuart et al. (2019). "Postreproductive killer whale grandmothers improve the survival of their grandoffspring." In: *Proceedings of the National Academy of Sciences* 116.52, pp. 26669–26673. ISSN: 0027-8424. DOI: 10.1073/pnas.1903844116. URL: https://www.pnas.org/content/116/52/26669.

Oppy, Graham and David Dowe (2019). "The Turing Test." In: *The Stanford Encyclopedia of Philosophy*. Ed. by Edward N. Zalta. Spring 2019. Metaphysics Research Lab, Stanford University.

Pajevic, Marija et al. (2018). "The relationship between the Dark Tetrad and a two-dimensional view of empathy." In: *Personality and Individual Differences* 123, pp. 125–130. ISSN: 0191-8869. DOI: https://doi.org/10.1016/j.paid.2017.11.009. URL: http://www.sciencedirect.com/science/article/pii/S0191886917306657.

Parton, A, P Malhotra, and M Husain (2004). "Hemispatial neglect." In: *Journal of Neurology, Neurosurgery & Psychiatry* 75.1, pp. 13–21. ISSN: 0022-3050. URL: https://jnnp.bmj.com/content/75/1/13.

Pearce, Eiluned, Christopher Stringer, and Robin Dunbar (2013). "New insights into differences in brain organization between Neanderthals and anatomically modern humans." In: *Proceedings. Biological sciences / The Royal Society* 280. DOI: 10.1098/rspb.2013.0168.

Pope, Sarah M. et al. (2020). "Optional-switch cognitive flexibility in primates: Chimpanzees' (Pan troglodytes) intermediate susceptibility to cognitive set." In: *Journal of Comparative Psychology* 134.1, pp. 98–109. DOI: 10.1037/com0000194. URL: https://doi.org/10.1037/com0000194.

Potts, Wayne K. and Edward K. Wakeland (1993). "Evolution of MHC genetic diversity: a tale of incest, pestilence and sexual preference." In: *Trends in Genetics* 9.12, pp. 408–412. ISSN: 0168-9525. DOI: https://doi.org/10.1016/0168-9525(93)90103-O. URL: http://www.sciencedirect.com/science/article/pii/016895259390103O.

Publishing, American Psychiatric (2013). *Diagnostic and statistical manual of mental disorders: DSM-5(tm), 5th ed.* Arlington, VA, US: American Psychiatric Publishing, Inc., pp. xliv, 947–xliv, 947. ISBN: 978-0-89042-555-8. DOI: 10.1176/appi.books.9780890425596. URL: https://www.psychologytoday.com/us/conditions/dissociative-identity-disorder-multiple-personality-disorder.

Ramachandran, V. S. (1998). "Consciousness and body image: lessons from phantom limbs, Capgras syndrome and pain asymbolia." In: *Philosophical Transactions of the Royal So-*

ciety of London. Series B: Biological Sciences 353.1377, pp. 1851–1859. DOI: 10.1098/rstb.
1998.0337. URL: https://royalsocietypublishing.org/doi/abs/10.1098/rstb.1998.0337.

— (2003). *The Emerging Mind*. Profile Books(GB); UK ed. edition. ISBN: 978-1861973030.

Ramachandran, V. S. and Sandra Blakeslee (1999). *Phantoms in the Brain: Probing the Mysteries of the Human Mind*. William Morrow Paperbacks. ISBN: 978-0688172176.

Ramachandran, V. S. and D. Rogers-Ramachandran (1996). "Synaesthesia in phantom limbs induced with mirrors." In: *Proc. R. Soc. Lond. B* 131, p. 263. DOI: 10.1098/rspb.1996.0058. URL: http://doi.org/10.1098/rspb.1996.0058.

Rand, Ayn, Harry Binswanger, and Leonard Peikoff (1990). *Introduction to Objectivist Epistemology*. Expanded 2nd ed. New American Library, New York, N.Y. ISBN: 04-5201-030-6.

Rathelot, Jean-Alban and Peter L. Strick (2009). "Subdivisions of primary motor cortex based on cortico-motoneuronal cells." In: *Proceedings of the National Academy of Sciences* 106.3, pp. 918–923. ISSN: 0027-8424. DOI: 10.1073/pnas.0808362106. URL: https://www.pnas.org/content/106/3/918.

Ravizza, Susan M. et al. (2011). "Left TPJ activity in verbal working memory: Implications for storage- and sensory-specific models of short term memory." In: *NeuroImage* 55.4, pp. 1836–1846. ISSN: 1053-8119. DOI: https://doi.org/10.1016/j.neuroimage.2010.12.021. URL: http://www.sciencedirect.com/science/article/pii/S1053811910016046.

Read, Dwight W. (2008). "Working Memory: A Cognitive Limit to Non-Human Primate Recursive Thinking Prior to Hominid Evolution." In: *Evolutionary Psychology* 6.4. DOI: 10.1177/147470490800600413. URL: https://doi.org/10.1177/147470490800600413.

Redgrave, P., T.J. Prescott, and K. Gurney (1999). "The basal ganglia: a vertebrate solution to the selection problem?" In: *Neuroscience* 89.4, pp. 1009–1023. ISSN: 0306-4522. DOI: https://doi.org/10.1016/S0306-4522(98)00319-4. URL: http://www.sciencedirect.com/science/article/pii/S0306452298003194.

Ridderinkhof, Anna et al. (2017). "Does mindfulness meditation increase empathy? An experiment." In: *Self and Identity* 16.3, pp. 251–269. DOI: 10.1080/15298868.2016.1269667. URL: https://doi.org/10.1080/15298868.2016.1269667.

Rogan, Michael T., Ursula V. Stäubli, and Joseph E. LeDoux (1997). "Fear conditioning induces associative long-term potentiation in the amygdala." In: *Nature* 390.6660, pp. 604–607. ISSN: 1476-4687. DOI: 10.1038/37601. URL: https://doi.org/10.1038/37601.

Rosenthal-von der Putten, Astrid M. et al. (2019). "Neural Mechanisms for Accepting and Rejecting Artificial Social Partners in the Uncanny Valley." In: *Journal of Neuroscience*. ISSN: 0270-6474. DOI: 10.1523/JNEUROSCI.2956-18.2019. URL: https://www.jneurosci.org/content/early/2019/07/01/JNEUROSCI.2956-18.2019.

Roth, Gerhard and Ursula Dicke (2005). "Evolution of the brain and intelligence." In: *Trends in Cognitive Sciences* 9.5, pp. 250–257. ISSN: 1364-6613. DOI: https://doi.org/10.1016/j.tics.2005.03.005. URL: http://www.sciencedirect.com/science/article/pii/S1364661305000823.

— (2012). "Evolution of the brain and intelligence in primates." In: *Evolution of the Primate Brain*. Ed. by Michel A. Hofman and Dean Falk. Vol. 195. Progress in Brain Research. Elsevier, pp. 413–430. DOI: https://doi.org/10.1016/B978-0-444-53860-4.00020-9. URL: http://www.sciencedirect.com/science/article/pii/B9780444538604000209.

Rowe, Timothy B., Thomas E. Macrini, and Zhe-Xi Luo (2011). "Fossil Evidence on Origin of the Mammalian Brain." In: *Science* 332.6032, pp. 955–957. ISSN: 0036-8075. DOI: 10.1126/science.1203117. URL: https://science.sciencemag.org/content/332/6032/955.

Samson, Dana et al. (2004). "Left temporoparietal junction is necessary for representing someone else's belief." In: *Nature Neuroscience* 7.5, pp. 499–500. ISSN: 1546-1726. DOI: 10.1038/nn1223. URL: https://doi.org/10.1038/nn1223.

Samsonovich, Alexei V and Giorgio A Ascoli (2005). "A simple neural network model of the hippocampus suggesting its pathfinding role in episodic memory retrieval." In: *Learn-*

ing and memory (Cold Spring Harbor, N.Y.) 12.2, pp. 193–208. ISSN: 1072-0502. DOI: 10. 1101/lm.85205. URL: https://europepmc.org/articles/PMC1074338.

Santiesteban, Idalmis et al. (2012). "Enhancing Social Ability by Stimulating Right Temporoparietal Junction." In: *Current Biology* 22.23, pp. 2274–2277. ISSN: 0960-9822. DOI: 10.1016/j.cub.2012.10.018. URL: https://doi.org/10.1016/j.cub.2012.10.018.

Schacter, Daniel L. et al. (2012). "The future of memory: remembering, imagining, and the brain." eng. In: *Neuron* 76.4, pp. 677–694. ISSN: 1097-4199. DOI: 10.1016/j.neuron.2012. 11.001. URL: https://pubmed.ncbi.nlm.nih.gov/23177955.

Schenk, Thomas and Robert D. McIntosh (2010). "Do we have independent visual streams for perception and action?" In: *Cognitive Neuroscience* 1.1. PMID: 24168245, pp. 52–62. DOI: 10.1080/17588920903388950. URL: https://doi.org/10.1080/17588920903388950.

Schuwerk, Tobias et al. (2016). "The rTPJ's overarching cognitive function in networks for attention and theory of mind." In: *Social Cognitive and Affective Neuroscience* 12.1, pp. 157–168. ISSN: 1749-5016. DOI: 10.1093/scan/nsw163. URL: https://doi.org/10. 1093/scan/nsw163.

Seinfeld, S. et al. (2018). "Offenders become the victim in virtual reality: impact of changing perspective in domestic violence." In: *Scientific Reports* 8. ISSN: 2045-2322. DOI: 10.1038/ s41598-018-19987-7. URL: https://doi.org/10.1038/s41598-018-19987-7.

Sevinc, Gunes et al. (2019). "Strengthened Hippocampal Circuits Underlie Enhanced Retrieval of Extinguished Fear Memories Following Mindfulness Training." In: *Biological Psychiatry* 86.9, pp. 693–702. ISSN: 0006-3223. DOI: 10.1016/j.biopsych.2019.05.017. URL: http://www.sciencedirect.com/science/article/pii/S0006322319314076.

Shallice, Tim et al. (1989). "The Origins of Utilization Behaviour." In: *Brain* 112.6, pp. 1587– 1598. ISSN: 0006-8950. DOI: 10.1093/brain/112.6.1587. URL: https://doi.org/10.1093/ brain/112.6.1587.

Shave, Robert E. et al. (2019). "Selection of endurance capabilities and the trade-off between pressure and volume in the evolution of the human heart." In: *Proceedings of the National Academy of Sciences* 116.40, pp. 19905–19910. ISSN: 0027-8424. DOI: 10.1073/pnas. 1906902116. URL: https://www.pnas.org/content/116/40/19905.

Sheth, Bhavin R. and Ryan Young (2016). "Two Visual Pathways in Primates Based on Sampling of Space: Exploitation and Exploration of Visual Information." In: *Frontiers in Integrative Neuroscience* 10, p. 37. ISSN: 1662-5145. DOI: 10.3389/fnint.2016.00037. URL: https://www.frontiersin.org/article/10.3389/fnint.2016.00037.

Silani, Giorgia et al. (2013). "Right Supramarginal Gyrus Is Crucial to Overcome Emotional Egocentricity Bias in Social Judgments." In: *Journal of Neuroscience* 33.39, pp. 15466– 15476. ISSN: 0270-6474. DOI: 10.1523/JNEUROSCI.1488-13.2013. URL: https://www. jneurosci.org/content/33/39/15466.

Skaggs, W. E. and B. L. McNaughton (1998). "Spatial firing properties of hippocampal CA1 populations in an environment containing two visually identical regions." eng. In: *The Journal of neuroscience : the official journal of the Society for Neuroscience* 18.20, pp. 8455–8466. ISSN: 0270-6474. DOI: 10.1523/JNEUROSCI.18-20-08455.1998. URL: https://pubmed.ncbi.nlm.nih.gov/9763488.

Sockol, Michael D., David A. Raichlen, and Herman Pontzer (2007). "Chimpanzee locomotor energetics and the origin of human bipedalism." In: *Proceedings of the National Academy of Sciences* 104.30, pp. 12265–12269. ISSN: 0027-8424. DOI: 10.1073/pnas.0703267104. URL: https://www.pnas.org/content/104/30/12265.

Soutschek, Alexander et al. (2016). "Brain stimulation reveals crucial role of overcoming self-centeredness in self-control." In: *Science Advances* 2.10. DOI: 10.1126/sciadv.1600992. URL: https://advances.sciencemag.org/content/2/10/e1600992.

Squair, Jordan and Jordan Squair (2012). *Craniopagus: Overview and the implications of sharing a brain.*

Squire, Larry R. (2009). "The legacy of patient H.M. for neuroscience." eng. In: *Neuron* 61.1, pp. 6–9. ISSN: 1097-4199. DOI: 10.1016/j.neuron.2008.12.023. URL: https://pubmed.ncbi.nlm.nih.gov/19146808.

Stoeckel, Cornelia et al. (2009). "Supramarginal gyrus involvement in visual word recognition." In: *Cortex* 45.9, pp. 1091–1096. ISSN: 0010-9452. DOI: https://doi.org/10.1016/j.cortex.2008.12.004. URL: http://www.sciencedirect.com/science/article/pii/S0010945209000264.

Stoerig, P and A Cowey (1997). "Blindsight in man and monkey." In: *Brain* 120.3, pp. 535–559. ISSN: 0006-8950. DOI: 10.1093/brain/120.3.535. URL: https://doi.org/10.1093/brain/120.3.535.

Striedter, Georg F., Shyam Srinivasan, and Edwin S. Monuki (2015). "Cortical Folding: When, Where, How, and Why?" In: *Annual Review of Neuroscience* 38.1. PMID: 25897870, pp. 291–307. DOI: 10.1146/annurev-neuro-071714-034128. URL: https://doi.org/10.1146/annurev-neuro-071714-034128.

Suchan, Boris et al. (2013). "Reduced connectivity between the left fusiform body area and the extrastriate body area in anorexia nervosa is associated with body image distortion." In: *Behavioural Brain Research* 241, pp. 80–85. ISSN: 0166-4328. DOI: https://doi.org/10.1016/j.bbr.2012.12.002. URL: http://www.sciencedirect.com/science/article/pii/S0166432812007760.

Tanaka, Shoji and Eiji Kirino (2019). "Increased Functional Connectivity of the Angular Gyrus During Imagined Music Performance." In: *Frontiers in Human Neuroscience* 13, p. 92. ISSN: 1662-5161. DOI: 10.3389/fnhum.2019.00092. URL: https://www.frontiersin.org/article/10.3389/fnhum.2019.00092.

Todd, J. Jay, Daryl Fougnie, and René Marois (2005). "Visual Short-Term Memory Load Suppresses Temporo-Parietal Junction Activity and Induces Inattentional Blindness." In: *Psychological Science* 16.12. PMID: 16313661, pp. 965–972. DOI: 10.1111/j.1467-9280.2005.01645.x. URL: https://doi.org/10.1111/j.1467-9280.2005.01645.x.

Tye, Kay M. et al. (2008). "Rapid strengthening of thalamo-amygdala synapses mediates cue-reward learning." In: *Nature* 453.7199, pp. 1253–1257. ISSN: 1476-4687. DOI: 10.1038/nature06963. URL: https://doi.org/10.1038/nature06963.

Versace, Elisabetta, Silvia Damini, and Gionata Stancher (2020). "Early preference for face-like stimuli in solitary species as revealed by tortoise hatchlings." In: *Proceedings of the National Academy of Sciences*. ISSN: 0027-8424. DOI: 10.1073/pnas.2011453117. URL: https://www.pnas.org/content/early/2020/09/09/2011453117.

Volz, Kirsten G., Ricarda I. Schubotz, and D. Yves von Cramon (2005). "Variants of uncertainty in decision-making and their neural correlates." In: *Brain Research Bulletin* 67.5. 2nd Conference on NeuroEconomics - ConNEcs 2004, pp. 403–412. ISSN: 0361-9230. DOI: https://doi.org/10.1016/j.brainresbull.2005.06.011. URL: http://www.sciencedirect.com/science/article/pii/S0361923005002273.

Walker, Alan (2009). "The Strength of Great Apes and the Speed of Humans." In: *Current Anthropology* 50.2, pp. 229–234. DOI: 10.1086/592023. URL: https://doi.org/10.1086/592023.

Watzek, Julia, Sarah M. Pope, and Sarah F. Brosnan (2019). "Capuchin and rhesus monkeys but not humans show cognitive flexibility in an optional-switch task." In: *Scientific Reports* 9.1, p. 13195. ISSN: 2045-2322. DOI: 10.1038/s41598-019-49658-0. URL: https://doi.org/10.1038/s41598-019-49658-0.

Webb, Taylor W., Hope H. Kean, and Michael Graziano (2016). "Effects of Awareness on the Control of Attention." In: *Journal of Cognitive Neuroscience* 28.6, pp. 842–851. DOI: 10.1162/jocn_a_00931. URL: https://doi.org/10.1162/jocn_a_00931.

Welford, Alan Traviss (1980). *Reaction times*. Academic Pr, pp. 73–128.

White, Thomas I. (2007). *In Defense of Dolphins: The New Moral Frontier*. Blackwell Publishing. ISBN: 978-14-0515-779-7.

Whitehead, Kimberley, Judith Meek, and Lorenzo Fabrizi (2018). "Developmental trajectory of movement-related cortical oscillations during active sleep in a cross-sectional cohort of pre-term and full-term human infants." In: *Scientific Reports* 8.17516. ISSN: 2045-2322. DOI: 10.1038/s41598-018-35850-1. URL: https://doi.org/10.1038/s41598-018-35850-1.

Wilkins, Jayne and Michael Chazan (2012). "Blade production 500 thousand years ago at Kathu Pan 1, South Africa: support for a multiple origins hypothesis for early Middle Pleistocene blade technologies." In: *Journal of Archaeological Science* 39.6, pp. 1883–1900. ISSN: 0305-4403. DOI: https://doi.org/10.1016/j.jas.2012.01.031. URL: http://www.sciencedirect.com/science/article/pii/S030544031200043X.

Wimmer, Heinz and Josef Perner (1983). "Beliefs about beliefs: Representation and constraining function of wrong beliefs in young children's understanding of deception." In: *Cognition* 13.1, pp. 103–128. ISSN: 0010-0277. DOI: https://doi.org/10.1016/0010-0277(83)90004-5. URL: http://www.sciencedirect.com/science/article/pii/0010027783900045.

Wolpert, Daniel M., Susan J. Goodbody, and Masud Husain (1998). "Maintaining internal representations: the role of the human superior parietal lobe." In: *Nature Neuroscience* 1.6, pp. 529–533. ISSN: 1546-1726. DOI: 10.1038/2245. URL: https://doi.org/10.1038/2245.

Wrangham, Richard and Johann Grolle (2019). *'Those Who Obeyed the Rules Were Favored by Evolution'*. [online; last accessed Apr 27th, 2019]. URL: http://www.spiegel.de/international/interview-with-anthropologist-richard-wrangham-a-1259252.html.

Young, Liane et al. (2010). "Disruption of the right temporoparietal junction with transcranial magnetic stimulation reduces the role of beliefs in moral judgment." In: *Proceedings of the National Academy of Sciences of the United States of America* 107, pp. 6753–6758. DOI: 10.1073/pnas.0914826107.

Yu, Feng et al. (2014). "A new case of complete primary cerebellar agenesis: clinical and imaging findings in a living patient." In: *Brain* 138.6, e353. ISSN: 0006-8950. DOI: 10.1093/brain/awu239. URL: https://doi.org/10.1093/brain/awu239.

Zanella, Matteo et al. (2019). "Dosage analysis of the 7q11.23 Williams region identifies BAZ1B as a major human gene patterning the modern human face and underlying self-domestication." In: *Science Advances* 5.12. DOI: 10.1126/sciadv.aaw7908. URL: https://advances.sciencemag.org/content/5/12/eaaw7908.

Index

7q11.23 duplication syndrome, 39

action
    prioritization, 18
    self-generated, 81
ADD, *see* attention deficit disorder
addiction
    therapy, 82
African elephant, 86
    cranial capacity, 30
afterlife
    immaterial, 103
akinetopsia, 63
al-Haitham, Ibn, 97
alcohol
    addiction, 82
alien hand syndrome, 70–73, 150
allegory of the cave, 102
allocortex, **7**
altruism, 91
amygdala, **18**, 58, 191
    connection, 134
    emotional mapping, 167
    learning, 20
    supervised learning, 166
    training, 167
angular gyrus, 36, **65**, 66, 77
    left, 64
    right, 64
animal
    intelligence, 30
anorexia nervosa, 56, 62
aphantasia, 155
aphasia, 83
apraxia, 63
Aquinas, Thomas, 103, 104
Aristotle, 102, 103
arthropod, 9, 31
artificial intelligence, 10
attention, **4**, 9, 11, 18, 126
    timeline, 5
    without awareness, 126, 137
attention deficit disorder, 153
attention schema, **179**, 180, 181, 188,
    193, 196
attention span, 154
Australopithecus, 29
    cranial capacity, 30

awareness, **126**
awareness schema, **188**, 190

backpropagation, 167, 170, 176
balance, 23
basal ganglia, **17**, 22, 69
    coordination, 152
BAZ1B, 39
Beagle, Peter S., xv, 223
bipedalism, 34
black box, 166
blind-deaf, 196
blindness
    inattentional, 156
blindsight, **124**, 125, 126, 137, 204
body
    injury, 73
    internal state, 64, 172
body schema, **62**, 69, 73, 172, 179–181,
    190
    baby, 69
    extension, 74, 172
    update, 74, 154
bottlenose dolphin, 29
bouba-kiki effect, 203
brain
    architecture, 7
    embryonic development, 32
    evolution, 8, 33
    folding, 32
    multiple, 32
    prediction, 67
    prediction machine, 49
    processing delay, 49
brain in a brain, 120
brain-to-body quotient, 32
Broca's area, **83**
bucket brigade, 167
Buddhism, **110**
    focus, 109

calcium signalling, 8
Calvin, William H., 115
Capgras delusion, **59**, 61
capuchin monkey, 40
Cartesian theater, **119**, 121, 197
cat, 86
    cranial capacity, 30

cephalopod, 31
cerebellum, **22**, 126
    balance, 23
    coordination, 152
    learning, 22
    missing, 23
cerebral cortex, **8**
cerebral hemispheres, **42**
cerebrum, **7**
checkers shadow illusion, 48
chess, 169, 186, 187
chicken-and-egg problem, 33
chimpanzee, 29
    aggressiveness, 38
    baby, 40
    cranial capacity, 30
    endurance, 34
    reaction time and memory, 37
Christianity, 103
classification
    images, 10
    signal, 9, 175
    XOR, 10
cognitive trade-off hypothesis, 36
color
    extra-spectral, 47
    spectral, 44
    wavelength, 46
    white, 47
complexity, 140
concept
    building, 54
    creation, 171
    mapping, 53
conceptualization, 54, 151
confabulation, 136
consciousness, 101, 133, **194**
    attribution, 93, 161
    emergent property, 123, 129, 138
    evolution, 195
    hard problem of, **197**
    illusion, 181
    language, 196
    level, 192, 195
    loop, 138, 140, 141, 188, 190
    model, 181
    multiple, 132
    origin, 106
    system, 123, 138
control
    feedback loop, 163
    muscle, 81
    overshooting, 163
    settling time, 163
    undershooting, 163
cooking, 34
coordination
    bilateral, 70
Copenhagen interpretation, **113**
coping, 81
corico-basal ganglia-thalamo-cortical
    loop, 159
corpus callosum, **127**
    damage, 129, 130, 132, 133, 158
cortical folding, 32
Cotard delusion, 60
cranial capacity, 30
Cretaceous–Paleogene extinction event, 24

daydreaming, 156
declarative memory, **146**
    transfer, 146
decussation, 128
delusion
    monothematic, **59**, 137
Dennett, Daniel, 119, 120
Descartes, René, 104, 105
dissociative identity disorder, 72
dog, 86
    cranial capacity, 30
dolphin, 29, 86
dorsal stream, 50, 52, 148
drawing, 10
dream, 21
dualism, **102**

edge detection, 10
egotism, 65
embryonic development, 32
emergent property, **121**
emotion, 78
    reading, 65
empathy, 88
    with self, 91
encephalization quotient, **30**, 31
endurance, 34
epileptic seizure, 129
episodic buffer, **144**
epistemology, 105
errata, xiii
Eurasian magpie, 86
    cranial capacity, 30

evil demon, **104**
evolution, 7, 115
    eye, 115
    influence of environment, 33
    niche, 33
    selection, 115
evolution of attention, 5
executive functions, **153**
experiences, 79
explaining, 87
extra-spectral color, 47
extrastriate body area, 56
eye
    damage, 123
    evolution, 115
    movement, 15

face recognition, 55
face-blindness, *see* prosopagnosia
facial recognition, 56
fallacy
    homunculus argument, **121**
fawn response, 19
feedback loop, 163
Feynman, Richard, 1, 229
fight, flight, fawn, freeze reaction, 25
fight, flight, freeze, fawn response, 19
filtering, 10
fire, 34
fish, 15
    thalamus, 16
flight response, 19
focus, 79, 188
    oscillation, 153
Form
    ideal, 104
fractal, 32
free will, 109
freeze response, 19
Fregoli delusion, 59
frontal lobe, **68**
fusiform body area, 56
fusiform face area, 55, 56, 58

gag reflex, 14
games, 37
garden path sentences, 67
gesticulation, 83
gesture
    interpretation, 65
giant manta ray, 86

gibbon, 29
Go, 187
goal, 188
gorilla, 29
    cranial capacity, 30
    encephalization quotient, 31
grammar, 66
Graziano, Michael, v
grip, 65
gustatory system, 14
gut feeling, 204
gyrus, **8**

habit, 82, 136
handedness, 127
hard problem of consciousness, **197**
hearing voices, 156
hemianopia, 134
hemispatial neglect, **134**, 134, 136, 137, 184, 204
    awareness, 134
hemisphere, 127
    communication, 72, 130
    differences, 127
    left, 53
    right, 53
    specialization, 128
    synchronization, 132, 158
    training, 132
Hering illusion, 49
hippocampus, **21**, 21, 53
    connection, 134
    evolution of, 20
    memory, 147
Hogen, Krista and Tatiana, 158, 159
Homo erectus, 29
    cranial capacity, 30
Homo habilis, 29
    cranial capacity, 30
Homo heidelbergensis, 29
Homo neanderthalensis, 29
Homo sapiens, 29
homunculus, 197
homunculus argument fallacy, **121**
hormonal system, 9
    suppressing function, 143
human
    baby, 41
    bipedalism, 34
    childhood, 40
    cranial capacity, 30

domestication, 38
    endurance, 34
    evolution, 33
    fire and cooking, 34
    hunting, 35
    intelligence, 33
    language, 36
    longevity, 41
    migration, 31
    modern, 29
    reaction time and memory, 36
    sight, 35
    task-switching, 40
    tool use, 38
hunting, 35
Hydra, 8, 27
hyperfocus, 153
hyperphantasia, 156

idealism, pluralistic, 109
image
    classification, 11, 175
    edge detection, 10
imagination, 155, 156, 176, 187
immaterial
    connection, 107
immune system, 13
impulsivity, 91
Indian idealism, **109**
infinity, 140
influence
    unconscious, 135
information
    flow, 193
inner ear, 23
inner voice, 158
intelligence
    evolution, 33
intention, 78
interneuron, 27
intuition, **97**
Islam, 103

janitor's dream, 115, 118
Jesus, 103
Judaism, 103
jumping spider, 31

Keller, Helen, 196
killer whale
    longevity, 41

lancelet, 11
language, 36, 196
    garden path sentences, 67
    grammar, 66
    tool use, 66
Last Unicorn, The, xv
lateral geniculate nucleus, **17**, 44, 47, 48,
    52, 124, 125, 139
law, 40
learning
    effect, 193
    neuronal, 167
Lee, Bruce, 125
Leibniz, Gottfried, 108, 109
LGN, *see* lateral geniculate nucleus
lie detection, 92
lobe, **42**
long-term memory, 21, 53, 146
    creation, 146
    encoding, 146
longevity, 41
loop of consciousness, **159**
lTPJ, *see* temporoparietal junction
lying, 92

macro-world, 115, 116, 118
macroscopic world, 117
Malebranche, Nicolas, 108
mammal
    domestication, 39
many-minds interpretation, **113**
mate selection, 13
materialism, **107**
measurements, 133
mechanistic theory, **138**
memory, 36, 79
    activation, 147
    false, 150
    hormones, 143
    long-term, 21, 53, 146
    loop, 143
    phonological, 66
    recall, 147
    regenerative capacitor, 143
    short-term, 53, 143
    spatial, 53
mental cinema, *see* hyperphantasia
MHC complex, 13
mind
    non-physical, 105
mind's eye, 155

mind-body dualism, *see* mind-brain
        dualism
mind-brain dualism, **102**, 103, 106, 107,
        129
mindfulness, 65
mirror
        smell, 86
        therapy, 73
mirror test, 85, **86**
mirrored-self misidentification, 60
model, **171**, 193
        attention, 180
        update, 172
model of attention, 193
Monism, **107**
monothematic delusion, **59**
motion detection, 10
motivation, 80
motor
        control, 27, 68
        coordination, 68
        program, 18, 22, 152
                competition, 70
multi-tasking
        overhead, 154
multiple personality disorder, *see*
                dissociative identity
                disorder
muscle, 27
        control, 27, 81
        memory, 125
mutation, 115

Neanderthal, 29, 31
        cranial capacity, 30
neocortex, **7**, 7, 12, 21, 23, 156
        evolution, 7
nervous system
        multi-layered, 10
Neumann-Wigner interpretation, **114**
neuron, 32, 167
neutral monism, **108**
New World monkeys, 26
Newton, Isaac, 47
nocturnal bottleneck hypothesis, **23**
non-verbal communication, 83

object permanence, 78, **79**, 84, 101, 172,
        179
Objectivism, **110**
        free will, 109

observer, 120
obsessive compulsive disorder, 80, 154
occipital lobe, **42**, 42, 63, 185
OCD, *see* obsessive compulsive disorder
octopus, 132
Old World monkeys, 26
olfactory system, 11–14
        loop, 140
        refinement, 13
ontology, 105
optic ataxia, 63
optic chiasm, 15, 42, 44
        damage, 123
optic tectum, *see* superior colliculus
optical illusion, 48, 49, 181
orangutan, 29
orientation, 20
out-of-body experience, 64, 75

pain
        location, 73
        mechanism, 167
parallelism, 108, 109
        psychophysical, 108
parietal lobe, **51**, 63, 64, 68, 185
        postcentral gyrus, 63
Pepper, 161
perception, 142
        self, 64
Permian-Triassic mass extinction event,
        23
perspective
        taking someone else's, 91
        understanding, 87
phantom limb pain, 74, 154
philosophical zombie, **197**
phonological loop, **144**
photoreceptor cell, 44
        combination, 46
pineal gland, 104
planning, 80
Plato, 102–104
pleasure
        mechanism, 167
Plesiadapis, 26
pluralistic idealism, 109
point of view, 190
post-traumatic stress disorder, 81
postcentral gyrus, 63
posterior parietal cortex, 63
posture

interpretation, 65
pre-processing
    visual, 10
prefrontal cortex, **25**, 85, 87, 88, 92, 136,
    185, 191
    conflict resolution, 82, 155
    executive functions, 152
    focus, 188
    goal, 188
    planning, 80
    suppression, 152, 153, 188
premotor cortex, **69**, 82, 85
primacy of consciousness, 103, 106
primary motor cortex, **26**, 26, 27, 68, 152
    size, 28
primary somatosensory cortex, **64**, 65
prioritization, 18
prism experiment, 47
processing delay, 49
property
    emergent, *see* emergent property
prosopagnosia, 55, 56
prosthetics, 75
psychophysical parallelism, 133
PTSD, *see* post-traumatic stress disorder

quantum computer, 118
quantum mechanics
    interpretation, 112
    observer, 121
    spooky action at a distance, 109
quantum mind, **111**, 115–118
quantum theory
    many-minds interpretation, **113**
    von Neumann-Wigner
      interpretation, **114**
quantum world, 115–118
Quran, 103

Ramachandran, Vilaynur, 132
Rand, Ayn, 110
ranged weapons, 35
reaction time, 36
reading, 36
reduplicative paramnesia, 60
reflex, 70
    eye, 52
    gag, 14
    head-turning, 15
    pupillary, 15
relationship

forming, 176
religion, 110
    Abrahamic, 103
resurrection, 103
retina, 47
reward delay, 79
rhesus macaques, 40
risk management, 80
ritual, 195
robot
    conscious, 198
    Pepper, 161
rTPJ, *see* temporoparietal junction
rubber hand illusion, **74**
rules and laws, 40

Sally and Ann task, 86
savannah hypothesis, **34**
selection, 115
self, 86, 196
    control, 79
    development, 185
    identification, 78
    perception, 64
    reflection, 175, 194
self-domestication, 38
self-generated action, 81
sense
    smell
      unconscious, 155
    touch, 63
settling time, 163
shark
    behavior, 22
short-term memory, 10, 36, 53
    impairment, 150
sight, 35
signal
    classification, 9
    degradation, 140, 144
simultanagnosia, 63
SMA, *see* supplementary motor area
smell
    communication, 12
    sense of, 12
SMG, *see* supramarginal gyrus
somatosensory cortex, 86
Sorites paradox, 122
soul, 103, 104
spectral color, 44
speech, 82

impairment, 82
intonation, 83
spine, 27
Spinoza, Baruch de, 108
split-brain syndrome, **130**, 130, 133, 159
sponge, 8
strategy, 80, 82
stream of consciousness, 157, 179
Stroop effect, 155
subconscious, 126, 204
subjective experience, 101, 198
    delusion, 197
    linearity, 159
    perception, 200
    sense perception, 204
subjective idealism, **109**
substance abuse, 91
sulcus, **8**
superior colliculus, 15, **16**, 23, 52, 139,
    161
superior parietal lobe, **64**, 69
supervised learning, **166**, 167
    downsides, 166
    limits, 170
supplementary motor area, 70
supramarginal gyrus, **66**, 77, 78, 85
    left, 128
sympathy, 88
syndrome of subjective doubles, 60
synesthesia, 203

task-switching, 40
taste
    learning, 14
temporal lobe, **51**, 53, 63
temporoparietal junction, **76**, 86–88, 91,
    92, 148, 151, 184, 187
    concept, 151
    damage, 92, 150, 193
    training, 90
thalamus, 12, **17**, 58, 156
    damage, 123, 156
    sharing, 158
Thatcher effect, 56
theory of forms, 102
theory of mind, **76**, 84, 87, 101, 172, 179,
    181
therapy
    violent offenders, 90
thermostat
    feedback loop, 164

programming, 162
thinking
    iterative, 187
three-dimensional data, 48, 125
timeline of attention, 5
tool use, 31, 36, 37, 65, 66
    evolution, 38
    language, 66
    model, 172
training
    dog, 167
tree
    fractal, 32
trial and error, 166, 173
    vs model, 176
Turing test, **168**, 169, 176
    memorization, 169
Turing, Alan, 168
twins, conjoined, 158

uncanny valley effect, 60, **61**
uncertainty, 82
unsupervised learning, **173**
utilization behavior, 79, 150, 153, 188
    suppression, 183

ventral stream, 50, 52, 53, 148
violent offenders
    therapy, 90
virtual reality, 75, 90
virtual world, 109
visual cortex, 42, 44, 85, 139
    damage, 124
    feedback loop, 140
    location, 52
    optimization, 52
visual pre-processing, 10
visual system
    motion detection, 10
visuo-spatial scratchpad, **144**, 144
vitamin C, 26

Watts, Alan, 95
weapon
    ranged, 35
Wernicke's area, **53**, 64, 66, 77, 82, 101
whale, 30
    cranial capacity, 30
white
    color, 47
    light, 181

Williams-Beuren syndrome, 39
word processing
        phonological, 66
        visual, 66
working memory, **144**, 149, 150, 179, 185, 187, 188, 193

Wrangham, Richard, 40

XOR, 10

An Important Final Note

Writers are not performance artists. While there are book signings and public readings, most writers (and readers) follow their passion alone in their writing spaces at home, in a café, in a library, at the beach, or at a mountain retreat.

What applause is for the musician, **reviews** *are for the writer.*

Books create a community among readers; you can share your thoughts among all those who will or have read this book.

Please leave a thoughtful, honest review and help me to create such a community on the platform on which you have acquired this book. What did you like, what can be improved? To whom would you recommend it?

Thank you, also in the name of all the other readers who will be better able to decide whether this book is right for them. A positive review will increase the reach of the book; a negative review will improve the quality of the next book. I welcome both!

"

If the human race develops an electronic nervous system, outside the bodies of individual people, thus giving us all one mind and one global body, this is almost precisely what has happened in the organization of cells which compose our own bodies. We have already done it. [...] If all this ends with the human race leaving no more trace of itself in the universe than a system of electronic patterns, why should that trouble us? For that is exactly what we are now!

—Alan Watts, The Book on the Taboo Against Knowing Who You Are